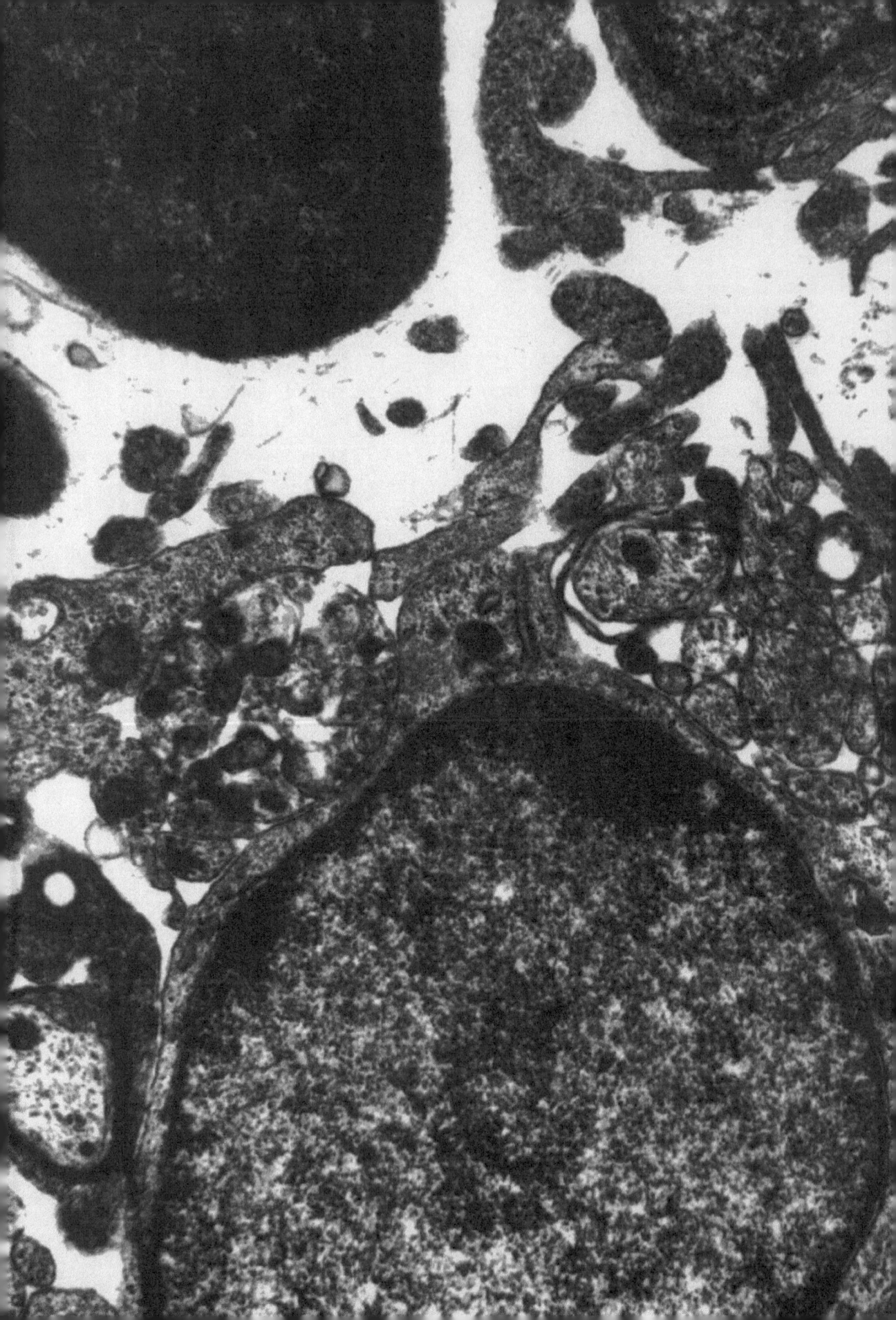

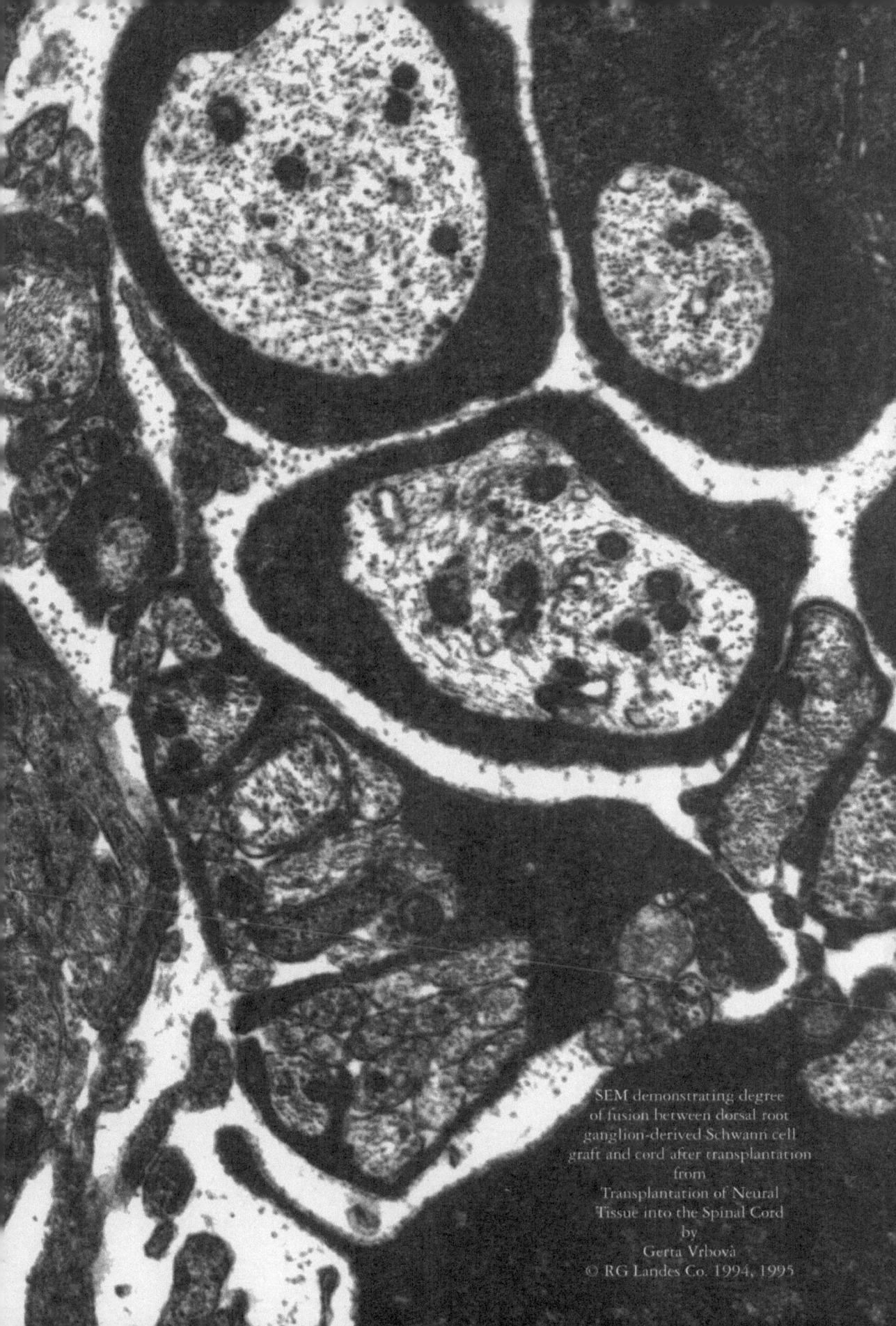

SEM demonstrating degree
of fusion between dorsal root
ganglion-derived Schwann cell
graft and cord after transplantation
from
Transplantation of Neural
Tissue into the Spinal Cord
by
Gerta Vrbová
© RG Landes Co. 1994, 1995

NEUROSCIENCE
INTELLIGENCE
UNIT

CALCIUM REGULATION BY CALCIUM-BINDING PROTEINS IN NEURODEGENERATIVE DISORDERS

NEUROSCIENCE
INTELLIGENCE
UNIT

CALCIUM REGULATION BY CALCIUM-BINDING PROTEINS IN NEURODEGENERATIVE DISORDERS

Claus W. Heizmann, Ph.D.

University of Zurich
Zurich, Switzerland

Katharina Braun, Ph.D.

Federal Institute for Neurobiology
Magdeburg, Germany

Springer-Verlag Berlin Heidelberg GmbH

NEUROSCIENCE INTELLIGENCE UNIT

CALCIUM REGULATION BY CALCIUM-BINDING PROTEINS IN NEURODEGENERATIVE DISORDERS

R.G. LANDES COMPANY
Austin, Texas, U.S.A.

Submitted: March 1995
Published: May 1995

International Copyright © 1995 Springer-Verlag Berlin Heidelberg
Originally published by Springer-Verlag, Heidelberg, Germany 1995
Softcover reprint of the hardcover 1st edition 1995

International ISBN 978-3-662-21691-0

Library of Congress Cataloging-in-Publication Data

Heizmann, Claus W.
    Calcium regulation by calcium-binding proteins in neurodegenerative disorders / by Claus W. Heizmann and Katharina Braun.
        p. cm. — (Neuroscience intelligence unit)
    Includes bibliographical references and index.
    ISBN 978-3-662-21691-0   ISBN 978-3-662-21689-7 (eBook)
    DOI 10.1007/978-3-662-21689-7

    1. Calcium-binding proteins—Physiological effect. 2. Calcium-binding proteins—Pathophysiology. 3. Neurochemistry. 4. Nervous system—Degeneration—Molecular aspects. I. Braun, Katharina. II. Title. III. Series.
        [DNLM: 1. Calcium-Binding Proteins—ultrastructure. 2. Calcium-Binding Proteins—phsiology. 3.Calcium—metabolism. 4.Central Nervous System Diseases—diagnosis. 5.Calcium-Binding Proteins—diagnostic use. QU 55 H473c 1995]
QP552.C24H45 1995
616.8 '047—dc20
DNLM/DLC   95-15197
for Library of Congress   CIP

PUBLISHER'S NOTE

R.G. Landes Company publishes five book series: *Medical Intelligence Unit, Molecular Biology Intelligence Unit, Neuroscience Intelligence Unit, Tissue Engineering Intelligence Unit* and *Biotechnology Intelligence Unit*. The authors of our books are acknowledged leaders in their fields and the topics are unique. Almost without exception, no other similar books exist on these topics.

Our goal is to publish books in important and rapidly changing areas of medicine for sophisticated researchers and clinicians. To achieve this goal, we have accelerated our publishing program to conform to the fast pace in which information grows in biomedical science. Most of our books are published within 90 to 120 days of receipt of the manuscript. We would like to thank our readers for their continuing interest and welcome any comments or suggestions they may have for future books.

Deborah Muir Molsberry
Publications Director
R.G. Landes Company

CONTENTS

PREFACE

In nerve cells, calcium ions activate and regulate a number of key processes, including fast axonal transport of substances, synthesis and release of some neurotransmitters, membrane excitability, and they might be involved in long-term potentiation and memory storage mechanisms. The Ca^{2+} message is transmitted into the intracellular response by Ca^{2+}-binding proteins that are involved in a variety of activities.

More than 10,000 articles on calcium were published in 1994 and more than 200 Ca^{2+}-binding proteins (characterized by a common structural motif, the EF-hand) have been discovered so far, emphasizing the tremendous interest and progress in Ca^{2+}-related research.

Impairment of Ca^{2+} homeostasis and an altered expression of Ca^{2+}-binding proteins (or deletions/mutations of the corresponding genes) have been implicated in acute insults such as stroke and epileptic seizures as well as neurodegenerative and bipolar affective disorders. Furthermore, Ca^{2+}-binding proteins have also been proven to be useful neuronal markers for a variety of functional brain systems and their circuitry.

This book summarizes the recent advances in the knowledge of the structure and physiological functions of the Ca^{2+}-binding proteins in the central nervous system, describes their cellular distribution and discusses their possible association with neurodegenerative disorders and their use as diagnostic tools.

In the future, Ca^{2+}-binding proteins might become important therapeutic targets for preventing neuronal death in several neurodegenerative disorders in man.

Acknowledgments

We would like to thank Mrs. Margrith Killen and Mrs. Ines Tiefert for typing and correcting the manuscript; Dr. A. Rowlerson for valuable suggestions; Dr. Beat Schäfer and Mr. Frank Neuheiser for help with preparation of the figures; and Dr. Martin Metzger and Mr. Wang Jizhong for help with preparing the confocal images.

=CHAPTER 1=

INTRODUCTION

Claus W. Heizmann and Katharina Braun

More than 10,000 articles were published in 1994 on calcium, emphasizing the widespread interest and progress in Ca^{2+}-related research. This book focuses mainly on Ca^{2+}-binding proteins in the central nervous system, where Ca^{2+} ions have been found to activate fundamental processes such as release of neurotransmitters, axonal flow, long term potentiation, cell motility, differentiation, secretion, and apoptosis. It has also been found that a number of neurodegenerative disorders have been attributed to aberrations of intracellular Ca^{2+} homeostasis.[1]

Intracellular Ca^{2+} levels and Ca^{2+} signaling within cells must be tightly controlled.[2,3] Ca^{2+} overload as a result of seizures or ischemia is supposed to activate biochemical processes, leading to enzymatic breakdown of proteins and lipids, malfunctioning of mitochondria, energy failure and ultimately cell death.[2] There is experimental evidence[3] that electrically induced irreversible depolarization of hippocampal neurons, which may be an early indication of neuronal damage, could be prevented by injecting Ca^{2+} chelators and thereby increasing intracellular buffering capacity. Thus, it is reasonable to assume that neurons containing certain intracellular Ca^{2+}-binding proteins, and therefore having a greater capacity to buffer Ca^{2+}, could be more resistant to degeneration.

Several research groups have now started to search for altered expressions of Ca^{2+}-binding proteins in affected brain regions of patients suffering from acute insults, such as stroke and epileptic seizures, and from chronic neurodegenerative disorders, such as Alzheimer's, Huntington's, Parkinson's and Pick's diseases.[1]

In addition, altered Ca^{2+} levels have been found in platelets of patients with bipolar affective disorders[4] and Ca^{2+} antagonists have been suggested for treatment of psychotic depression (see chapter 6).

Therefore, studies to advance our knowledge of the structure and physiological functions of these proteins and the neurological systems in which they are expressed may become important for the development of therapies to prevent neuronal death in an array of neurodegenerative diseases.

Furthermore, Ca^{2+}-binding proteins have been proven to be useful neuronal markers for a variety of functional brain systems and their circuitries.

In addition, Ca^{2+}-binding proteins may soon become valuable diagnostic tools and selective markers to estimate the extent of brain damage in various neurological disorders when measured in the cerebrospinal fluid or blood and also for the immunohistochemical classification of brain tumors in children and adults.

Ca^{2+}-coupled responses consist of three major steps: 1) Activating ligands such as hormones, growth factors, etc. bind to membrane receptors, resulting in a rise in intracellular calcium concentration; 2) Calcium will then bind to intracellular mediator proteins (EF-hand Ca^{2+}-binding proteins or annexins), which transmit the signal by modifying, in turn, specific target proteins; 3) These altered target proteins coordinate the cellular response to the stimulus. It is clear that the concentration of calcium ions inside the cell needs to be precisely controlled in a narrow range between resting and activating levels. This suggests that the proteins that bind calcium and therefore are involved in the regulation of the ion concentration are important in many biochemical regulatory processes (Fig. 1.1). A crucial step towards understanding the calcium-evoked responses, therefore, is the identification and characterization of internal Ca^{2+}-binding proteins. Various strategies such as isolation of high-affinity Ca^{2+}-binding proteins or sequence homology studies have led to the identification of a large number of proteins that may be involved in the calcium response.

In general, two families of proteins, the EF-hand Ca^{2+}-binding proteins and the annexins, Ca^{2+}-dependent and phospholipid-binding proteins, are involved in the cellular response.

Figure 1.1 also lists some extracellular proteins that are characterized by different consensus motifs for Ca^{2+}-binding than

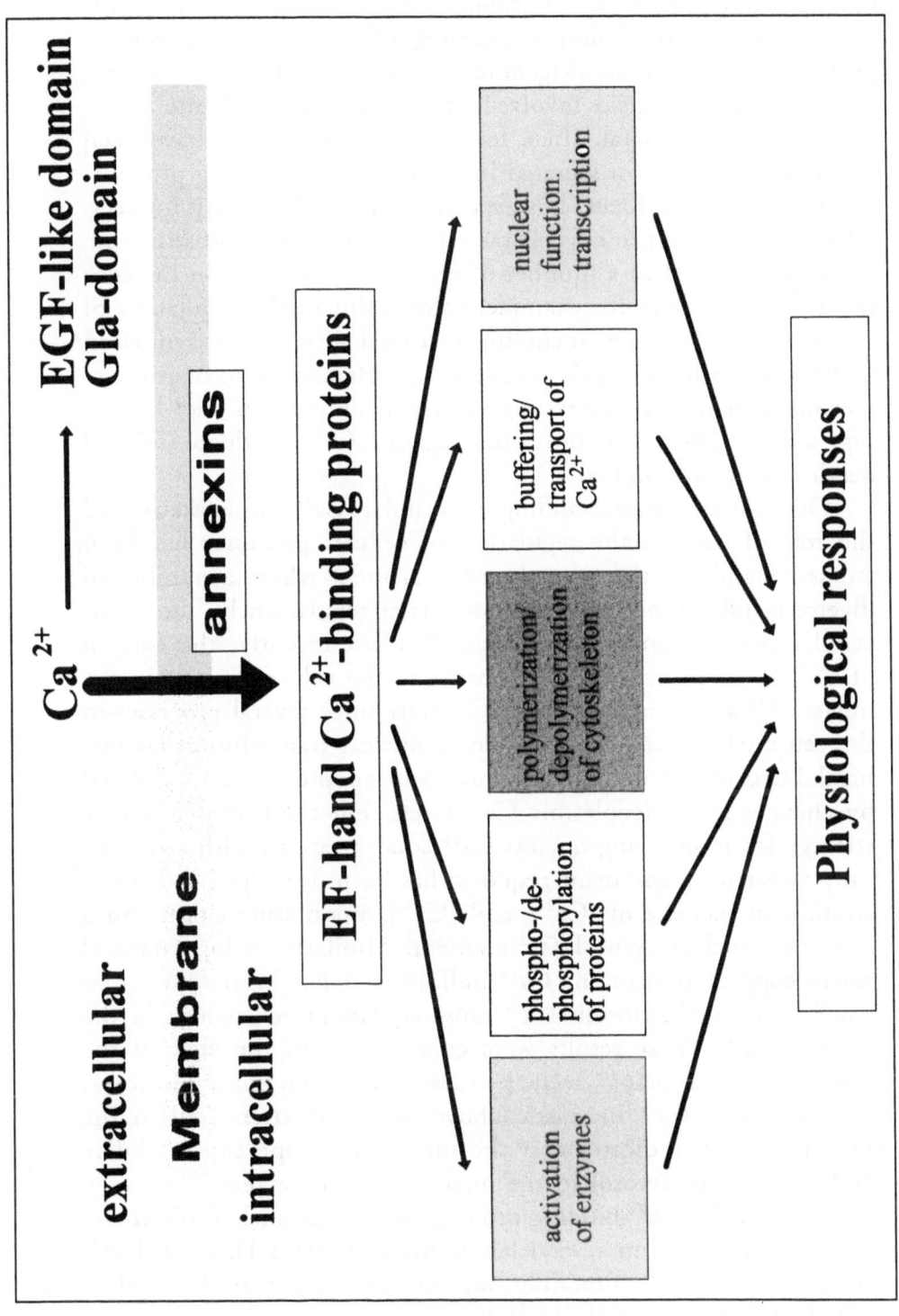

Fig. 1.1

the EF-hand proteins or annexins. These are (a) the matrix proteins fibrillin[5] and fibullin[6] containing EGF-like Ca^{2+}-binding repeats and (b) the non-collagenous protein, osteocalcin, containing γ-carboxy-glutamic acid involved in Ca^{2+}-binding and interaction with the bone mineral.[7] In a few cases these proteins have been found to be targets for secreted intracellular Ca^{2+}-binding proteins.

This book will focus, however, on those high-affinity intracellular Ca^{2+}-binding proteins characterized by the EF-hand structure. They are involved in a number of important functions in the central nervous system; for example, calmodulin in the regulation of many enzyme activities including neuronal nitric oxide synthase;[8] S100 proteins in cell cycle progression, differentiation, neurite extension, and tumor progression; and parvalbumin in Ca^{2+}-buffering. and transport and protection against Ca^{2+} overload and cell death (chapters 2 and 5).

Our current understanding of cellular Ca^{2+} homeostasis and the role of Ca^{2+} in the regulation of cellular processes has been studied mainly in the cytosolic compartment where a number of diverse cellular functions are under strict regulation by the intracellular free-calcium concentration.[9,10] Most recently, the control of Ca^{2+} levels in other compartments of the cell, especially in the nucleus[11,12] also became of major interest since several processes in the cell nucleus such as activation of nuclear transcription factors, breakdown of the nuclear envelope, and apoptosis are modulated by changes in nucleoplasmic Ca^{2+} levels. For these studies a new strategy for monitoring nuclear Ca^{2+} concentrations with a specifically targeted recombinant aequorin has been developed,[13] demonstrating an increase of $[Ca^{2+}]_i$ and $[Ca^{2+}]_n$ upon stimulation and a transient nucleus-cytosol Ca^{2+}gradient. Similarly, using confocal microscopy of fluorescent Ca^{2+} indicators it has been found that small changes of cytosolic Ca^{2+} cause equally rapid changes in the nuclear Ca^{2+}. These results were consistent with the view of the nuclear pore as being freely permeable to small ions. However, large cytosolic Ca^{2+} increases (above 300 nM) attenuated in the nucleus, which indicates that the nuclear envelope can modulate Ca^{2+} entry from cytosol to the nucleus to some extent.[14]

A search for Ca^{2+}-binding proteins in the cell nucleus has therefore been initiated in several laboratories (Table 1.1). High levels of calmodulin and calmodulin-binding proteins were detected in the nuclei of brain cells.[11,12,15] Very recently, the first report ap-

Table 1.1. Suggested functions of some EF-hand calcium-binding proteins in the cell nucleus

		selected references
Calmodulin	inhibition of transcription	1
	regulation of p68 RNA helicase	2
Calcineurin	activation of transcription	3,4
Nucleobindin	DNA-binding	5
	triggering DNA fragmentation in apoptosis?	
NEFA*	DNA-binding domain	6
	two helix-loop-helix (HLH) domains with two EF-hand motifs; leucine zipper domain; anti-DNA autoimmunization?	

*DNA binding/EF-hand/acidic amino acid rich protein

1. Corneliussen B, Holm M, Waltersson Y et al. Nature 1994; 368:760-764.
2. Buelt MK, Glidden BJ, Storm DR. J Biol Chem 1994; 269:29367-29370.
3. Tsuboi A, Muramatsu M, Tsutsumi A et al. Biochem Biophys Res Commun 1994; 199:1064-1072.
4. Jain J, McCaffrey PG, Miner Z et al. Nature 1993; 365:352-355.
5. Miura K, Kurosawa Y, Kanai Y. Biochem Biophys Res Commun 1994; 199:1388-1393.
6. Barnikol-Watanabe S, Gross NA, Götz H et al. Biol Chem Hoppe-Seyler 1994; 375:497-512.

peared describing the function of calmodulin in the nucleus.[16] Calmodulin was shown to bind to the basic helix-loop-helix domains (bHLH) of some transcription factors, resulting in the inhibition of their DNA binding. This study has identified a system whereby Ca^{2+} signals can directly regulate gene expression through selective Ca^{2+}-dependent calmodulin inhibition of bHLH DNA-binding domains.

Another example of a Ca^{2+}-binding protein acting in the cell nucleus is calcineurin (a Ca^{2+}-/calmodulin-dependent phosphatase), which is present in the central nervous system and was found to activate transcription.[17]

One of the most challenging future questions will be to study the structure, function, and mechanisms of action and interaction of Ca^{2+}-binding proteins in the cell nucleus.

The role of mitochondria in cell Ca^{2+} homeostasis has been reinvestigated. It has been demonstrated[18] that agonist-stimulated elevations of cytosolic free Ca^{2+} results in a rapid and transient increase of mitochondrial Ca^{2+}. Using a novel technique of specifically targeted recombinant aequorin[18] now offers the possibility to

monitor Ca^{2+} concentrations in a number of other cellular compartments (e.g. nucleus or endoplasmic reticulum) during cell activation.

Generally it is assumed that high-affinity Ca^{2+}-binding proteins
with the structural EF-hand motif exert their signaling function
mostly intracellularly since extracellularly their Ca^{2+}-binding sites
would always be saturated. However more recently, secretion of
several Ca^{2+}-binding proteins has been reported, extracellular binding proteins have been identified, and their possible extracellular
functions were discussed (chapter 5). Table 1.2 lists those EF-hand
proteins for which an extracellular function has been reported.
S100β, for example, is secreted by clonal glioma cell lines and by
primary rat cortical astrocytes in the nervous system.[19,20] S100β,
when secreted from glial cells, acts directly on neurons as a neurite
extension factor.[21] The mechanism of secretion is not yet known.
None of these proteins has a classical leader peptide that would
direct the protein to the secretory pathway. However, the routing
and secretion might be via membrane-bound proteins (e.g. members of the annexin family).

Recently, the extracellular 36-kD microfibril-associated glycoprotein was found to be the specific target for the intracellular
Ca^{2+}-binding protein, $S100A_4$ (or calvasculin), which is secreted
from smooth muscle cells,[22] supporting the view of its role in the
extracellular compartment.

The discovery of EF-hand Ca^{2+}-binding motifs in extracellular
multidomain proteins such as BM40/SPARC/osteonectin[23,24] or
integrins[25] is the best proof that proteins of the EF-hand protein
family exist and function in the extracellular space where calcium
levels are orders of magnitude higher than in the cytosol.

Members of the annexin family have been found to be exposed on the cell surface or even secreted from cells binding to
extracellular matrix proteins, e.g. annexin II to collagen type II
and X[26] or tenascin-C.[27] The mechanisms of secretion of annexins
and EF-hand Ca^{2+}-binding proteins is still not known but there is
growing evidence for their extracellular location (bound to matrix
proteins) and involvement in signal transduction.

Another area of interest is the binding of metal ions other than
Ca^{2+} to EF-hand proteins. A specific binding of Zn^{2+} to some members of Ca^{2+}-binding EF-hand proteins has been reported, e.g. to
$S100A_6$ (calcyclin),[28] present in a subpopulation of neurons in the
hippocampus.[29] This region of the brain, which is severely affected

Table 1.2. Suggested extracellular functioning of intracellular EF-hand calcium-binding proteins

BM40/SPARC/osteonectin:	extracellular matrix protein[1]
nucleobindin:	secreted protein; DNA binding properties[2,3]
S100β-dimer:	secreted from glioma cells, neuritic extension factor[4,5]
S100A8 (MRP8), **S100A9** (MRP-14):	secreted from myeloid cells[6]
S100A4 (CAPL, calvasculin):	secreted from smooth muscle cells and tumor cells; binding to an extracellular protein (MAP, Mr = 36kDa)[7]
β-parvalbumin (Avian Thymic Hormone, ATH):	thymic hormone activity (lymphocyte maturation)[8]
calmodulin:	development of preimplantation human embryos[9]

1. Maurer P, Mayer U, Bruch M et al. Eur J Biochem 1992; 205:233-240.
2. Everitt EA, Sage EH. Biochem Cell Biol 1992; 70:1368-1379.
3. Miura K, Kurosawa Y, Kanai Y. Biochem Biophys Res Commun 1994; 199:1388-1393.
4. Hilt DC, Kligman D. In: Heizmann CW, ed. Novel Calcium-Binding Proteins. Berlin Heidelberg: Springer-Verlag, 1991; 65-103.
5. Zimmer DB, van Eldik LJ. J Neurochem 1988; 50:572-579.
6. Hessian PA, Edgeworth J, Hogg N. J Leukocyte Biol 1993; 53:197-204.
7. Watanabe Y, Usuda N, Tsugane S et al. J Biol Chem 1992; 267:17136-17140.
8. Brewer JM, Wunderlich JK, Ragland W. Biochimie 1990; 72:653-660.
9. Woodward BJ, Lenton EA, McNeil S. Human Reproduction 1993; 8:272-276.

by lesions in senile dementia and Alzheimer's disease, is known to accumulate Zn^{2+}.[30] The functional significance of these events is not clear but an involvement of Ca^{2+}/Zn^{2+}-binding proteins might be considered.

Another area of interest is the binding of Mg^{2+} to some Ca^{2+}-binding proteins.[31] For example, parvalbumin (abundant in GABAergic neurons of the central nervous system) is a Ca^{2+}- and Mg^{2+}-binding protein thought to be involved in Ca^{2+} buffering/transport and protecting cells against Ca^{2+} overload (chapter 5). Generally, it has been believed that intracellular free Mg^{2+} concentrations do not vary in cells. Recently, however, intracellular free Mg^{2+} concentrations were measured in single rat brain neurons with the Mg^{2+}-sensitive fluorescent dye magnafura-2.[32] It was shown that glutamate increased intracellular free Mg^{2+}, thereby influencing neuronal excitability[32] and possibly the function of the cytosolic parvalbumin.

These observations indicate that alterations in Ca^{2+} homeostasis due to (a) an altered expression of Ca^{2+}-binding proteins, or (b) mutations in the Ca^{2+}- and Mg^{2+}-binding sites or sites of intra- or extracellular interactions might be involved in some neurodegenerative disorders in man.

The altered processing of β-amyloid precursor protein is probably responsible for the disruption of neuronal Ca^{2+} homeostasis and cell death in Alzheimer's disease.[33]

Other studies have shown that various forms of central nervous system injuries can lead to prominent changes in gene expression of Ca^{2+}-binding proteins such as calbindin D-28K as well as growth factors, stress proteins, and structural proteins.[34] This indicates that a perturbed Ca^{2+} homeostasis is associated with acute CNS injury and offers a possibility for therapeutic interventions.

Furthermore, rapid changes of expression of the calbindin D-28K gene in the cerebellum were recently found following morphine injection.[35] This suggested that an altered Ca^{2+} homeostasis may play an important role in acute opiate pharmacology and in the adaptation of the CNS to morphine administration.

There are also several examples where mutations have been detected in extra- and intracellular Ca^{2+}-binding proteins affecting Ca^{2+} homeostasis and causing disease. For example, mutations in the human Ca^{2+}-sensing receptor gene cause familial hypocalciuric hypercalcemia and neonatal severe hyperparathyroidism.[36]

In extracellular fibrillin (FBN_1), mutations in the EGF-like Ca^{2+}-binding domain (different from the EF-hand motif and Ca^{2+}-binding sites in annexins), which lead to an altered Ca^{2+}-binding, are involved in the pathogenesis of Marfan syndrome, a common autosomal dominant disorder of the extracellular matrix with cardinal manifestations in the ocular, skeletal, and cardiovascular systems.[5]

Another example is a mutation adjacent to the first Ca^{2+}-binding domain of platelet glycoprotein IIB, which may be responsible for Glanzmann thrombasthenia.[37]

Furthermore, mutations in the putative EF-hand Ca^{2+}-binding domain of polyomavirus VP1 affects the process of viral assembly[38] and further studies will provide important information about mechanisms of infectivity and hemagglutination.

Ca^{2+}-binding proteins are also tested as targets for pharmacological interventions.[39-41] For example, the activation of calpain I, a Ca^{2+}-dependent EF-hand thiol protease, has been implicated as a final and common event leading towards cell death following stroke, traumatic brain injury, spinal cord injury and several other forms of neurodegeneration. Novel inhibitors of calpain have been developed that are potent, selective, stable, and membrane permeant. When tested in in vivo animal models they provided substantial protection against focal ischemic brain damage even when administered hours after the ischemic event.[42,43]

Another target for therapeutic intervention is calmodulin, an EF-hand Ca^{2+}-binding protein regulating fundamental physiological processes. Several calmodulin antagonists have been developed with a high potency and selectivity with possible applications in vasodilatation.[39,40]

The need to control and prevent cancer has led investigators to develop cancer therapeutics with novel mechanisms of action. One such agent, carboxyamido-triazole (CAI), has recently entered phase I clinical trials. The mechanism of action of this drug is through signal transduction pathways and alterations in calcium homeostasis and opens new approaches to cancer therapy.[41]

As summarized in this Introduction, aberrations in the second messenger function of calcium are found in many diseases of the brain and other tissues, and have often been attributed to altered levels of calcium or Ca^{2+}-binding proteins. More than 200 of such proteins are known and are involved in intracellular functions from Ca^{2+}-buffering and transport, to regulation of enzymes, differentiation and secretion as well as extracellular functions from promoting neurite extension to hormone-like activities.

This book will survey their structures, interactions, tissue-specific expression and localization in the central nervous system, and their use as diagnostic tools and targets for therapeutic intervention.

REFERENCES
1. Heizmann CW, Braun K. Changes in Ca^{2+}-binding proteins in human neurodegenerative disorders. Trends Neurosci 1992; 15:259-264.
2. Siesjö BK, Wieloch T. In: Delgado-Escueta AV, Ward AA, Jr, eds. Advances in Neurology. Raven Press, 1986:813-847.
3. Scharfman HE, Schwartzkroin PA. Protection of dentate hilar cells from prolonged stimulation by intracellular calcium chelation. Science 1989; 246:257-260.

4. Dubovsky SL, Lee C, Christiano J et al. Lithium lowers platelet intracellular ion concentration in bipolar patients. Lithium 1991; 2:167-174.
5. Dietz HC, McIntosh I, Sakai LY et al. Four novel FBN1 mutations: significance for mutant transcript level and EGF-like domain calcium binding in the pathogenesis of Marfan syndrome. Genomics 1993; 17:468-475.
6. Pan T-C, Sasaki T, Zhang R-Z et al. Structure and expression of fibulin-2, a novel extracellular matrix protein with multiple EGF-like repeats and consensus motifs for calcium binding. J Cell Biol 1993; 123:1269-1277.
7. Theofan G, Haberstroh LM, Price PA. Molecular structure of the rat bone Gla protein gene and identification of putative regulatory elements. DNA 1989; 8:213-221.
8. Matsuoka A, Stuehr DJ, Olsoni JS et al. L-Arginine and calmodulin regulation of the heme iron reactivity in neuronal nitric oxide synthase. J Biol Chem 1994; 269:20335-20339.
9. Stuenkel EL. Regulation of intracellular calcium and calcium buffering properties of rat isolated neurohypophysial nerve endings. J Physiol 1994; 481.2:251-271.
10. Berridge MJ, Dupont G. Spatial and temporal signaling by calcium. Current Opinion Cell Biol 1994; 6:267-274.
11. Hoek JB, Thomas AP, eds. Calcium and the nucleus. Cell Calcium 1994; 16:237-338.
12. Gilchrist JSC, Czubryt MP, Pierce GN. Calcium and calcium-binding proteins in the nucleus. Mol Cell Biochem 1994; 135:79-88.
13. Brini M, Murgia M, Pasti L et al. Nuclear Ca^{2+} concentration measured with specifically targeted recombinant aequorin. EMBO J 1993; 12:4813-4819.
14. Al-Mohanna FA, Caddy KWT, Bolsover SR. The nucleus is insulated from large cytosolic calcium ion changes. Nature 1994; 367:745-750.
15. Vendrell M, Aligué R, Bachs O et al. Presence of calmodulin and calmodulin-binding proteins in the nuclei of brain cells. J Neurochem 1991; 57:622-628.
16. Corneliussen B, Holm M, Waltersson Y et al. Calcium/calmodulin inhibition of basic-helix-loop-helix transcription factor domains. Nature 1994; 368:760-764.
17. Tsuboi A, Muramatsu M, Tsutsumi A et al. Calcineurin activates transcription from the GM-CSF promoter in synergy with either protein kinase C or NF-κB/AP-1 in T cells. Biochem Biophys Res Commun 1994; 199:1064-1072.
18. Rizzuto R, Simpson AWW, Brini M et al. Rapid changes of mitochondrial Ca^{2+} revealed by specifically targeted recombinant aequorin. Nature 1992; 358:325-327.
19. Shashoua VE, Hesse GW, Moore BW. Proteins of the brain extra-

cellular fluid: evidence for release of S-100 protein. J Neurochem 1984; 42:1536-1541.

20. Zimmer DB, van Eldik LJ. Levels and distribution of the calcium-modulated proteins S-100 and calmodulin in rat C6 glioma cells. J Neurochem 1988; 50:572-579.

21. Hilt DC, Kligman D. The S-100 protein family: A biochemical and functional overview. In: Heizmann CW, ed. Novel Calcium-Binding Proteins. Berlin Heidelberg: Springer-Verlag, 1991:65-103.

22. Watanabe Y, Usuda N, Tsugane S et al. Calvasculin, an encoded protein from mRNA termed pEL-98, 18A2, 42A, or p9Ka, is secreted by smooth muscle cells in culture and exhibits Ca^{2+}-dependent binding to 36-kDa microfibril-associated glycoprotein. J Biol Chem 1992; 267:17136-17140.

23. Maurer P, Mayer U, Bruch M et al. High-affinity and low-affinity calcium binding and stability of the multidomain extracellular 40kDa basement membrane glycoprotein (BM-40/SPARC/osteonectin). Eur J Biochem 1992; 205:233-240.

24. Everitt EA, Sage EH. Overexpression of SPARC in stably transfected F9 cells mediates attachment and spreading in Ca^{2+}-deficient medium. Biochem Cell Biol 1992; 70:1368-1379.

25. Hogg N, Landis RC, Bates PA et al. The sticking point: how integrins bind to their ligands. Trends Cell Biol. 1994; 4:379-383.

26. Kirsch T, Pfaffle M. Selective binding of anchorin CII (annexin V) to type II and X collagen and to chondrocalcin (C-propeptide of type II collagen). FEBS Lett 1992; 310:143-147.

27. Chung CY, Erickson HP. Cell surface annexin II is a high affinity receptor for the alternatively spliced segment of tenascin-C. J Cell Biol 1994; 125:539-548.

28. Filipek A, Heizmann CW, Kuznicki J. Calcyclin is a calcium- and zinc-binding protein. FEBS-Lett 1990; 264:263-266.

29. Filipek A, Puzianowaska M, Cieslak B et al. Calcyclin - Ca^{2+}-binding protein homologous to glial S-100β is present in neurons. Neuro Report 1993; 4:383-386.

30. Vener AV, Aksenova MV, Burbaeva GS. Drastic reduction of the zinc- and magnesium-stimulated protein tyrosine kinase activities in Alzheimer's disease hippocampus. FEBS 1993; 328:6-8.

31. Williams RJP. Meeting report: Calcium-binding proteins in normal and transformed cells. Cell Calcium 1994; 16:339-346.

32. Brocard JB, Rajdev S, Reynolds IJ. Glutamate-induced increases in intracellular free Mg^{2+} in cultured cortical neurons. Neuron 1993; 11:751-757.

33. Mattson MP, Barger SW, Cheng B et al. β-Amyloid precursor protein metabolites and loss of neuronal Ca^{2+} homeostasis in Alzheimer's disease. Trends Neurosci 1993; 16:409-414.

34. Lowenstein DH, Gwinn RP, Seren MS et al. Increased expression of mRNA encoding calbindin-D28k, the glucose-regulated proteins,

or the 72 kDa heat-shock protein in three models of acute injury. Brain Res Mol Brain Res 1994; 22:299-308.

35. Tirumalai PS, Howells RD. Regulation of calbindin D28k gene expression in response to acute and chronic morphine administration. Mol Brain Res 1994; 23:144-150.

36. Pollak MR, Brown EM, Wu Chou Y et al. Mutations in the human Ca^{2+}-sensing receptor gene cause familial hypocalciuric hypercalcemia and neonatal severe hyperparathyroidism. Cell 1993; 75:1297-1303.

37. Poncz M, Rifat S, Coller BS et al. Glanzmann thrombasthenia secondary to a Gly273 -> Asp mutation adjacent to the first calcium-binding domain of platelet glycoprotein IIb. J Clin Invest 1994; 93:172-179.

38. Haynes JI, Chang D, Consigli RA. Mutations in the putative calcium-binding domain of polyomavirus VP1 affect capsid assembly. J Virology 1993; 67:2486-2495.

39. Hidaka H, Tanaka T. Transmembrane Ca^{2+} signaling and a new class of inhibitors. Meth Enzymol 1987; 139:570-582.

40. Mannhold R, Caldirola P, Bijloo GJ et al. New calmodulin antagonists of the diphenylalkylamine type. I. Biological activity, SAR and the role of lipophilicity. Eur J Med Chem 1993; 28:727-734.

41. Cole K, Koh E. Calcium-mediated signal transduction: biology, biochemistry, and therapy. Cancer and Metastasis Rev 1994; 13:31-44.

42. Hong S-C, Goto Y, Lanzino G et al. Neuroprotection with a calpain inhibitor in a model of focal cerebral ischemia. Stroke 1994; 25:663-669.

43. Bartus RT, Baker KL, Heiser AD et al. Postischemic administration of AK275, a calpain inhibitor, provides substantial protection against focal ischemic brain damage. J Cerebral Blood Flow Metab 1994; 14:537-544.

CHAPTER 2

STRUCTURES OF EF-HAND CA²⁺-BINDING PROTEINS AND ANNEXINS

The calcium signal is transduced into an intracellular response in part by calcium-binding proteins are thought to be involved in the regulation of many cellular activities. These proteins may be subdivided into two groups with distinct structural features:

(A) The calcium modulated proteins, including calmodulin, troponin-C, parvalbumins, S100 proteins and calbindin D-28K (Tables 2.1 and 2.2). Which share a structural feature, the EF-hand (Fig. 2.1, top).

(B) The annexin protein family, whose members interact with phospholipids and cellular membranes in a calcium-dependent manner. These proteins show a high degree of sequence homology and share a repetitive conserved sequence. This segment of 70 amino acid residues includes the site for calcium binding and for association with phospholipids (Fig. 2.1, bottom). A common nomenclature has recently been proposed for these proteins, which have previously been known as lipocortins, calcimedins, calpactins, or chromobindins (for nomenclature see Table 2.3). Their localization and functions are listed in Table 2.4.

In this review we will focus mainly on the high-affinity, calcium-binding proteins characterized by their EF-hand structure, summarizing their protein (this chapter) and gene (chapter 3) structures, distribution, localization, developmental appearance (chapter 4), physiological functions (chapter 5) and use in the diagnosis of neurodegenerative disorders (chapter 6). Many new members

Table 2.1. EF-hand Ca²⁺-binding proteins in the brain

Proteins with an EF-hand structural motif	Localization	Suggested functions	Neurodegenerative disorder	Selected references
1. calmodulin CaMI, CaMII	ubiquitous, large projection neurons	mediates many Ca^{2+}-dependent processes	Keshina, Alzheimer's disease	1
2. parvalbumin	subpopulation of GABAergic neurons	Ca^{2+}-buffering and transport; protective role against Ca^{2+}-overload; transmitter release	Alzheimer's disease, epilepsy, ischemia, Pick's disease, Down's syndrome, meningiomas neurofibromatosis	2-13
calbindin D28K	neurons (expression regulated by corticosterone in hippocampus)	Ca^{2+}-buffering and transport; protective role against Ca^{2+} overload	Alzheimer's disease, epilepsy, ischemia, Parkinson's disease Down's syndrome ALS schizophrenia	11, 13-19
calretinin	neurons	Ca^{2+}buffering and transport; phosphorylation	Alzheimer's disease schizophrenia	19-23
3. recoverin, visinin, frequinin, p26, VILIP, S-modulin, hippocalcin, GCAP, CBP-18, chromogranin A, NVP's, neurocolcin, NCS	photoreceptor layer in retina, several regions of the brain, periglomerular cells and dendrites in olfactory bulb; pyramidal cells in hippocampus, neurons	phototransduction, activates guanylate cyclase to restore dark state inhibition of rhodopsin kinase	cancer-associated retinopathy Alzheimer's disease	24-31 46-48
4. calcineurin	neurons	calmodulin-dependent phosphatase; target of cyclophilin-cyclosporin complexes	?	32, 44
5. calpain	all neurons, astro- and microglia	Ca^{2+} activated protease	Alzheimer's disease ischemia	5,33
6. 90-kDa diacyl-glycerol kinase	preferentially neurons	signal transduction	?	34

7. ERC-55 (ER Ca²⁺-binding protein of 55 kDa)	in ER	distinct subclass of subfamily of EF-hand proteins; functions unknown	?	35
reticulo-calbin (homologous to ERC-55)	in ER	as above	?	36
8. S100A₁ (S-100α)	neurons	assembly/disassembly of microtubules, phosphorylation	?	37
S-100β, S100A₆ (calcyclin)	glia cells, brain, other tissues	growth/differentiation factor; regulates intracellular Ca²⁺ concentration; assembly and disassembly of microtubules and actin filaments; phosphorylation; extracellular function: neurite extension, serotonergic growth factor in vitro	Alzheimer's disease, Down's syndrome, AIDS	38-40
9. R₂D₂ antigen	olfactory receptor neurons	a Ca²⁺-binding phosphoprotein; modulation of olfactory signal transmission	?	41
calcyphosine (p24)	brain, thyroidea	?	?	42
CCBP-23 (homologous to R₂D₂ and calcyphosine)	muscle, brain	?	?	43

1. McLachlan DRC, Song L, Bergeron C et al. Alzheimer Disease and Associated Disorders. 1987; 1:1781-1790.
2. Arai H, Emson PC, Mountjoy CQ et al. Brain Res 1987; 418:164-169.
3. Hof PR, Morrison JH. Exp Neurol 1991; 111:293-301.
4. Brady DR, Mufson EJ. Soc Neurosci Abstr 1991; 17:691.
5. Iwamoto H, Emson PC. Neurosci Lett 1991; 128:81-84.
6. Satoh J, Tabira T, Sano M et al. Acta Neuropathol 1991; 81:388-395.
7. Kamphuis W, Huisman E, Wadman WJ et al. Brain Res 1989; 479:23-34.
8. Sloviter RS, Sollas AL, Barbaro NM et al. J Comp Neurol 1991; 308:381-396.
9. Sloviter RS. Science 1987; 235:73-75.
10. Vonau M, Törk I. Soc Neurosci Abstr 1991; 17:1260.
11. Johansen FF, Tonder N, Zimmer J et al. Neurosci Lett 1990; 120:171-174.
12. Arai H, Noguchi I, Makino Y et al. J Neurol 1991; 238:200-202.
13. Kobayashi K, Emson PC, Mountjoy CQ et al. Neurosci Lett 1990; 113:17-22.
14. Ichimiya Y, Emson PC, Mountjoy CQ et al. Brain Res 1988; 475:156-159.
15. Lowenstein DH, Miles MF, Hatam F et al. Neuron 1991; 6:627-633.
16. Sonnenberg JL, Frantz GD, Lee SH et al. Mol Brain Res 1991; 9:179-190.
17. Yamada T, McGeer PL, Baimbridge KC et al. Brain Res 1990; 526:303-307.
18. Ferrer I, Tunon T, Serrano MT et al. J Neurol Neurosurg Psychiatry 1993; 56:257-261.
19. Davis SR, Lewis DA. Soc Neurosci Abstr 1993; 19:84-89.
20. Brion JP, Résibois A. Acta Neuropathol 1994; 88:33-43.
21. Hof PR, Nimchinsky EA, Celio MR et al. Neurosci Lett 1993; 152:145-149.
22. Rogers JH. J Cell Biol 1987; 105:1343-1353.
23. Winsky L, Yamaguchi T, Jacobowitz DM. Soc Neurosci Abstr 1991; 17:605.
24. Stryer L. J Biol Chem 1991; 266:10711-10714.
25. Lambrecht H-G, Koch K-W. EMBO J 1991; 10:793-798.
26. Lenz SE, Braun K, Braunewell KH et al. J Neurochem 1994; 63:72.
27. Cox JA, Durussel I, Compte M et al. J Biol Chem 1994; 269:32807-32813.
28. Polans AS, Burton MD, Haley TL et al. Investigative Ophthal & Visual Scie 1993; 34:8190.
29. Saitoh S, Takamatsu K, Kobayashi M et al. Neurosci Lett 1993; 157:107-110.
30. Adams LA, Munoz DG. Acta Neuropathol 1993; 86:365-370.
31. Lipp H-P, Wolfer DP, Qin WX et al. J Neurochem 1993; 60:1639-1649; Addendum: J. Neurochem 1993; 61:79.
32. Goto S, Matsukado Y, Uemura S et al. Exp Brain Res 1988; 69:645-650.
33. Perlmutter LS, Gall C, Baudry M et al. J Comp Neurol 1990; 296:269-276.
34. Goto K, Kondo H. Proc Natl Acad Sci USA 1993; 90:7598-7602
35. Sambrook JF. Cell 1990; 61:197-199.
36. Weis K, Griffiths G, Lamond AI. J Biol Chem 1994; 269:19142-19150.
37. Hilt DC, Kligman D. In: Heizmann CW, ed. Novel Calcium Binding Proteins. Fundamentals and Clinical Implications. Berling: Springer Verlag, 1991:65-105.
38. Griffin WS, Stanley LC, Ling C et al. Proc Natl Acad Sci USA 1989; 86:7611-7615.
39. Kato K, Suzuki F, Kurobe N et al. J Molec Neurosci 1990; 2:109-113.
40. Allore R, O'Hanlon D, Proce R et al. Science 1988; 239:1311-1313.
41. Nemoto Y, Ikeda J, Katoh K et al. J Cell Biol 1993; 123:963-976.
42. Lefort A, Lecocq R, Libert F et al. EMBO J 1989; 8:111-116.
43. Sauter A, Staudenmann W, Hughes GJ et al. Eur J Biochem 1995;227:97-101.
44. Mulkey R M, Endo S, Shenolikar S et al. Nature 1994; 369:486-487.
45. Kordowska J, Apel A, Brand I et al. Acta Histochem. Cytochem. 1994; 27:205-218.
46. Cox JA, Durussel I, Comte M et al. J Biol Chem 1994; 269:32807-32813
47. Kawamura S, Cox JA, Nef D. Biochem Biophys Res Commun 1994; 203:121-127.
48. Ladant D. J Biol Chem 1995; 270:3179-3185.

Table 2.2. Current nomenclature for s100 genes not clustered

New nomenclature for S100 genes clustered on 1q21

previous name	previous symbol	synonyms	new name *	new symbol*
S100 alpha	S100A	S100α, S100	S100 calcium-binding protein A1	S100A1
S100L	S100L	CaN19	S100 calcium-binding protein A2	S100A2
S100E	S100E		S100 calcium-binding protein A3	S100A3
Murine placental calcium protein	CAPL	p9Ka, 42A, pEL98, mts1, metastasin, calvasculin, 18A2	S100 calcium-binding protein A4	S100A4
S100D	S100D		S100 calcium-binding protein A5	S100A5
Calcyclin	CACY	2A9, PRA, CaBP, 5B10	S100 calcium-binding protein A6	S100A6
Psoriasin	PSOR1		S100 calcium-binding protein A7	S100A7
Calgranulin A	CAGA	CFAg, MRP8, p8, MAC 387, 60B8Ag, L1Ag, CP-10, MIF, NIF	S100 calcium-binding protein A8	S100A8
Calgranulin B	CAGB	CFAg, MRP14, p14, MAC 387, 60B8Ag, L1Ag, MIF, NIF	S100 calcium-binding protein A9	S100A9
Calpactin light chain	CAL1L	CLP11, p11, Cal[1], p10, 42C	S100 calcium-binding protein A10	S100A10

Current nomenclature for S100 genes not clustered

name	symbol	synonyms	chromosome
S100 calcium-binding protein beta	S100B	S100β, S100 beta, NEF, S100	21q22
Calbindin 3	CALB3	CaBP9k, ICaBP, Calbindin-D9k	Xp22
S100 calcium-binding protein P	S100P		4p16

* New nomenclature for S100 genes located at 1q21 as discussed at the recent "Third European Symposium on Calcium Binding Proteins in Normal and Transformed Cells" held in Zürich, March 6-9, 1994 and approved by the human nomenclature committee, April 29, 1994. All names and symbols listed (except for synonyms and old symbols/names) correspond to the approved and official entries in the Genome Database (GDB).

Reprinted with permission from: Genomics 25. 1995; 638-643 © Academic Press.

of this protein family have been discovered recently (those expressed in the brain are listed in Table 2.1), bringing the total to over 200 proteins. Calcium-binding proteins with known functions, such as calmodulin, troponin-C, myosin light chains, calpain, or calcineurin, are far outnumbered by those whose roles are not known. Most of them are expressed in a cell-type specific manner. Recently, this family of proteins has attracted additional interest, since altered concentrations of some of these proteins have been reported in disease states of the central nervous system. An exploration of their structures and involvement in mechanisms of calcium-mediated regulation in normal cells should therefore help to elucidate the underlying mechanisms of these pathological conditions.

A. EF-HAND CA²⁺-BINDING PROTEINS

Calmodulin, troponin-C, parvalbumin, S1OO proteins, and many new members of this family exhibit a common structural motif, the EF-hand, present in multiple copies, which bind calcium selectively and with high affinity[1,2] (Fig. 2.1, top). Each of these domains consists of a loop of 12 amino acids (a variant loop with 14 amino acids is present in the S1OO protein subfamily) that is flanked by two α-helices. This helix-loop-helix Ca²⁺-binding motif was first identified with the crystal structure of the calcium-binding carp parvalbumin and is designated as EF-hand after the E- and F-helices of parvalbumin.[1,3] The prototype loop consists of 12 amino acids of which 5 have a carboxyl (or a hydroxyl group) in their side chain and are precisely spaced so as to coordinate the calcium ion. The sixth ligand is a water molecule.

The universality of this structural motif was confirmed by the analysis of the crystal structure of further members, i.e., the vitamin D-dependent intestinal calcium-binding protein calbindin-9K, (a member of the S100 subfamily), troponin-C, and calmodulin (for review see[2]). Using this structural information a putative EF-hand dependent calcium-binding ability of a protein can be predicted or confirmed on the basis of the amino acid or cDNA sequence, leading to the discovery of many new EF-hand proteins. The model even allows to predict whether an EF-hand domain is still functional or if it has lost its calcium-binding ability due to mutations.

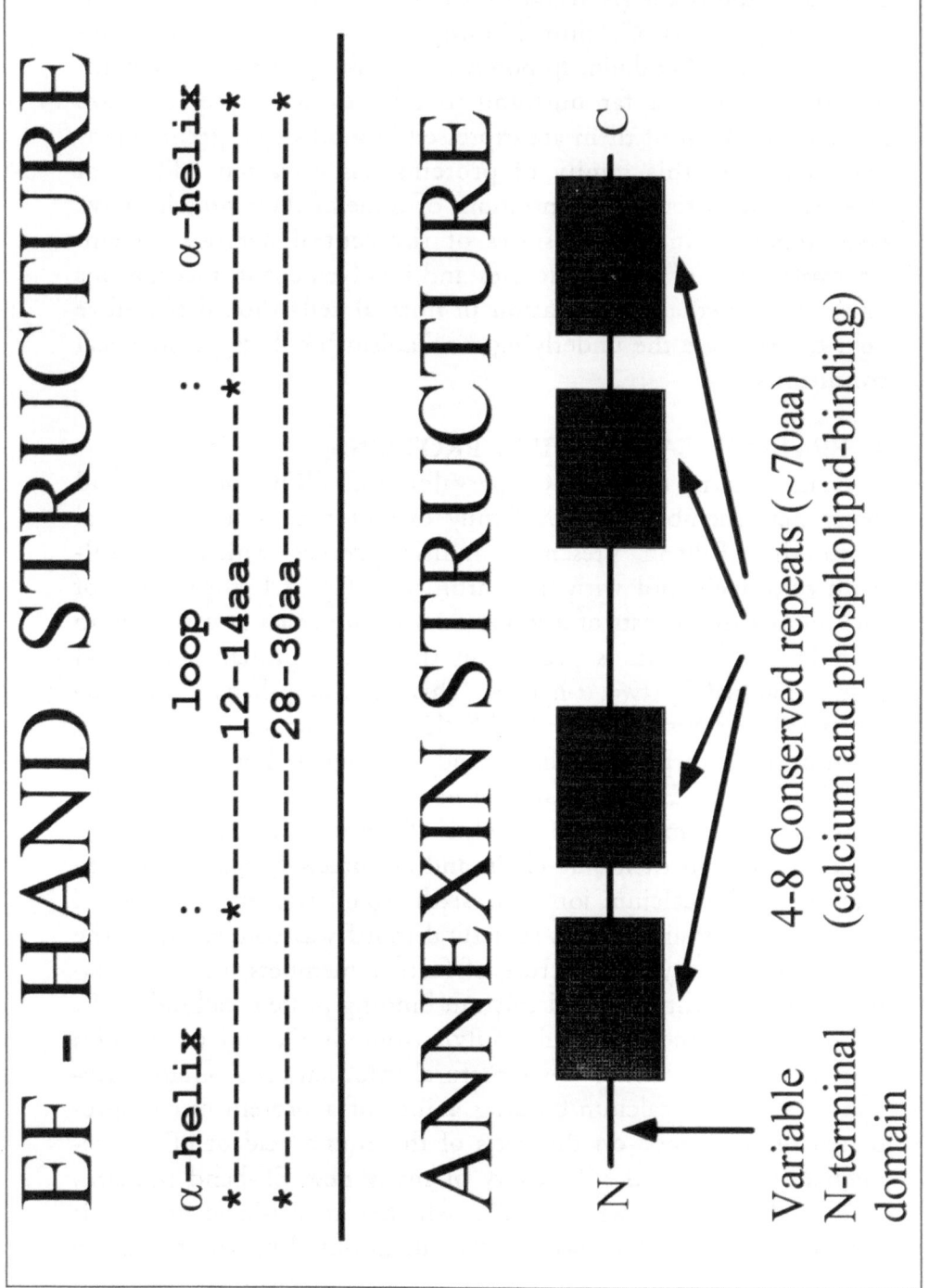

Fig. 2.1

Table 2.3. Nomenclature of the annexin family

Annexin	Synonyms
I	Lipocortin I
	Calpactin II
	p35
	Chromobindin 9
II	Calpactin I heavy chain
	Lipocortin II
	p36
	Chromobindin 8
	Protein I
	Placental anticoagulant protein IV
III	Lipocortin III
	Placental anticoagulant protein III
	35-α calcimedin
IV	Endonexin I
	Protein II
	32.5 kDa calelectrin
	Lipocortin IV
	Chromobindin 4
	Placental anticoagulant protein II
	Placental protein 4-X
	35-β calcimedin
V	Placental anticoagulant protein I
	Inhibitor of blood coagulation
	Lipocortin V
	35 kDa calelectrin
	Endonexin II
	Placental protein 4
	Vascular anticoagulant-β
	35-γ calcimedin
	Calphobindin I
	Anchorin CII
VI	p68, p70, 73k
	67 kDa calelectrin
	Lipocortin VI
	Protein III
	Chromobindin 20
	67 kDa calcimedin
	Calphobindin II
VII	Synexin
VIII	Vascular anticoagulant-β
IX and X	(only reported in Drosophila)
XI	
XII	(only reported in Hydra)
XIII	ISA, intestinal-specific annexin

Most known EF-hand proteins contain multiple copies (two to eight) of the helix-loop-helix domain, generally arranged in pairs so that two domains have contact with each other in pairs, which seems to enhance the affinity for calcium. In calbindin-9K the two EF-hands are packed against each other to yield a globular structure. The pair of functional domains in parvalbumin has a similar structure and the third degenerated amino terminal domain is also packed against them, maintaining a globular structure. The known 4-domain proteins (calmodulin and troponin-C) fold into two calbindin-D9K-like structures connected by a long central α-helix that gives these molecules a dumbbell-like appearance.

The X-ray crystallographic data available so far have been obtained with calcium bound to the sites. However, circular dichroism studies show that major conformational changes occur on addition of calcium to the apoproteins. These conformational changes are considered to be responsible for the calcium-dependent interaction with EF-hand acceptor proteins.

Table 2.1 lists those EF-hand Ca^{2+}-binding proteins expressed in the central nervous system. They can be subdivided into the following subfamilies:

1. CALMODULIN

Calmodulin, containing four Ca^{2+}-binding sites, is a multifunctional protein and responsible for the activation of many enzymes, such as type II CaM kinase, calcineurin, adenylcyclase and phosphodiesterase.[4,5] Calmodulin is involved in several neuronal functions, e.g. activation of these enzymes and release of neurotransmitters.[6] Calmodulin is ubiquitously distributed and is particularly abundant in testis and brain.[5-8] In mammals, three non-allelic genes encode a single calmodulin molecule by synonymous codon usage. Three calmodulin genes of the rat, *CaMI*, *CaMII* and *CAMIII* are strongly expressed in the brain.[7]

The specific distribution of the three calmodulin gene RNAs was investigated in the brain by in situ hybridization. Their suggested functions and possible involvement in neurodegenerative disorders are listed in Table 2.1.

A strong hybridization signal was obtained in the large projection neurons of the CNS; the molecular mechanisms of the neuron-specific expression of the calmodulin genes are currently under investigation in several laboratories.

2. CYTOSOLIC PROTEINS.

The second group of EF-hand Ca^{2+}-binding proteins (listed in Table 2.1) are the cytosolic proteins, parvalbumin, calbindin D-28K, and calretinin. These proteins are of special interest because of their suggested excito-protective roles in neurons and their association with several diseases of the brain and apoptosis (chapters 5 and 6).

Parvalbumins are Ca^{2+}-binding proteins of the EF-hand type.[1,2] Alpha- and β-parvalbumins are generally recognized in lower and higher vertebrates, including man.[2,9,10] They can be distinguished by their different isoelectric points (α>5, β<5), sequence characteristics, affinities for Ca^{2+} and Mg^{2+},[11] crystal structures (α-parvalbumin;[12] β-parvalbumin[3,13,14]) cell-type specific expression[9,15-19] and chromosomal localization (Table 3.1).

Whereas lower vertebrates have up to five isoforms of parvalbumin,[10] mammals including man have two parvalbumins (an α- and a β-form),[19,20] mainly present in a subpopulation of GABAergic neurons of the brain (chapter 4).

Beta-parvalbumin (also named oncomodulin) is not expressed in brain tissue but is restricted to human placenta.[21] In several tissues the pattern of α- and β-parvalbumin expression is different in man than in other vertebrates.[19-22] For example, rat fast-twitch skeletal muscles contain high concentrations of α-parvalbumin (involved in muscle relaxation) whereas, in a comparable human muscle, α-parvalbumin is restricted to the muscle spindles.[19] The species- and cell-type specific expression has implications for functional studies which will be discussed later (chapter 5).

Calbindin D-28K is present in all vertebrate species and in a wide range of different cells within these species, as well as in some invertebrates. Intestines of non-mammalian vertebrates as well as kidney and cerebellum of all vertebrates contain the highest calbindin D-28K concentrations. Calbindin D-28K synthesis in intestine, kidney and pancreas is regulated by 1,25-dihydroxyvitamin D_3, but in other tissues, including brain, the expression of the protein seems to be independent of this hormone.[23] The protein may function as a mediator of vitamin D-dependent Ca^{2+} transport, and a kinetic model has been proposed for calbindin as an intracellular facilitator of intestinal or renal calcium diffusion.[24]

The distribution and function of calbindin D-28K in the central nervous system has been investigated by several groups (chapters 4 and 5).

The amino acid sequence of calbindin D-28K from several species has been deduced from the analysis of cDNA or genomic clones[25-29] or by Edman degradation.[30-32] The protein has a calculated molecular weight of 30 kD and contains six putative calcium-binding sites (EF-hand domains). Due to evolutionary mutations, domains II and VI have probably lost their calcium-binding ability.[33] Therefore calbindin D-28K binds only 3-4 calcium/molecule with an affinity of $K_a > 10^6$.[34]

A gene encoding the neuronal protein calretinin has been isolated as a cDNA clone from chick retina. Calretinin (Mr = 29kD) is structurally homologous to calbindin D-28K but its tissue distribution is quite different. mRNAs from both genes are abundant in the retina and in many areas of the brain, but calretinin mRNA is absent in intestine and other non-neuronal tissues. Calretinin and calbindin D-28K are expressed in different sets of neurons throughout the brain (chapter 4). In view of the close homology and mutually distinct localization it is suggested that both proteins may perform similar functions in different sets of neurons.[35,36] The function of calretinin is presently under investigation, using, for example, the hammerhead ribozyme approach[37] to destroy designated mRNAs in living cells, which permits the study of gene function by producing null phenotypes.

3. RECOVERIN-LIKE PROTEINS

Recoverin is one member of a new EF-hand subfamily[38] also including S-modulin, a protein from frog rod outer segments,[39] visinin from retinal cones,[40] neurocalcins (also named hippocalcin, visinin-like protein, VILIP, recoverin-like protein)[41-46] immunologically related to CBP-18[47] localized in special regions of the brain, or frequinin,[48] which is expressed in *Drosophila* synapses.

The 3-dimensional structure of recoverin has recently been elucidated.[49] This protein contains four EF-hands arranged in a compact array different from the dumbbell shape of the two other 4-EF-hand proteins, calmodulin and troponin-C.

The human recoverin gene is a 9-10 Kb single-copy gene with three exons and two introns mapped to chromosome 17.[50,51] Recoverin is specifically expressed in the human retina, both in the photoreceptors and in a subpopulation of bipolar cells.[52]

Because of its potential role in photoreceptor physiology, recoverin has been considered as a candidate gene for *retinitis pigmentosa,* and experiments are in progress to test this possibility. Retinitis pigmentosa is a group of inherited retinal degeneration diseases characterized by progressive loss of peripheral vision and night blindness as a result of rod photoreceptor cell death.

Recently, recoverin (or a closely related protein, but not visinin) was identified as the auto-antigen recognized by the sera of patients affected with a cancer-associated retinopathy.[51,53-55] In this disease the development of a tumor at a distant site leads to the degeneration of photoreceptors.

The function of recoverin is presently unknown. The previous view was that recoverin stimulates the photoreceptor guanylate cyclase at submicromolar Ca^{2+} concentrations, however, this conclusion has recently been retracted.[56]

4. CALCINEURIN

Calcineurin is a protein phosphatase (previously named protein phosphatase 2B) that requires Ca^{2+} for enzyme activation. Calcineurin is a heterodimer composed of a 59-61 kD calmodulin-binding catalytic subunit (calcineurin A) and a 19 kD calcium-binding regulatory subunit (calcineurin B) with four EF-hand Ca^{2+}-binding sites similarly to calmodulin.[57,58] A distinct cellular expression of calcineurin isoenzyme was found in the brain.[59] Ca^{2+} can stimulate calcineurin activity by binding either directly to calcineurin B or to calmodulin.

To date three distinct genes have been cloned for calcineurin B in mammals. Two of these three genes (α and β form) are highly expressed in brain. The expression of the third (γ) calcineurin B gene is testis-specific.[60-62]

The physiological role of this enzyme in the brain is only beginning to be understood but its high levels suggest it must have an important function there.

Recently it has been found that calcineurin is a molecular target for the immunosuppressants, cyclosporin A and FK506, suggesting that the inhibition of the phosphatase activity is the mechanism by which the immunosuppressant exerts its effects.[63,64]

The distinct expression of two isoforms of calcineurin in the rat brain suggests specific functions in modulating neuronal activity in the particular cell types.[65] This view received some recent support[66] when it was found that dynamin I (a terminal phosphoprotein with intrinsic GTPase activity required for exocytosis) was dephosphorylated by calcineurin; this inhibited the GTPase activity of dynamin I in vitro and after depolarization of nerve terminals. The effect in nerve terminals was blocked by the calcineurin inhibitor cyclosporin A. It was therefore concluded that calcineurin represents a Ca^{2+} sensitive switch for depolarization in evoked synaptic vesicle recycling.

Calcineurin has also been found to be associated with long-term depression, an activity-dependent decrease in synaptic efficacy and a mechanism permitting neural networks to store information more effectively.[67]

5. Calpain

The Ca^{2+}-dependent neutral protease, calpain, consists of one large (80 kD) and one small (30 kD) subunit. In mammalian tissues, including brain, two isozymes exist with distinct calcium sensitivities. The 80 kD catalytic subunit contains four distinct domains. There is a marked similarity of the second domain with the papain-like thiol proteases. The fourth domain located at the C-terminal end is homologous to calmodulin and contains four EF-hand motifs. This indicates that calpain arose from the fusion of genes for thiol protease and EF-hand Ca^{2+}-binding protein. So far there have been only a few reports of gene fusion of proteins with different functions and evolutionary origin.[68-74]

The 30 kD subunit is composed of two domains. The N-terminal domain is rich in glycine. The C-terminal end is also a domain homologous to domain IV of the 80 kD subunit and to calmodulin.

Calpain isoforms I and II have been implicated in several aspects of brain function including neurofilament turnover, Wallerian degeneration, and excitatory synaptic transmission.

Immunohistochemical studies have shown that calpain I is located mainly in neurons, whereas calpain II was found to be more prominent in glial cells.[75]

The presence of calpain II in senile plaques suggested an involvement in the pathogenesis of Alzheimer's disease.[76-78] It was suggested that pharmacological modulation of the calpain system may be considered for a potential therapeutic strategy, including the newly discovered tissue-specific calpain species.[79]

6. DIACYLGLYCEROL KINASE

The 90-kD enzyme is involved in signal transduction converting the second messenger diacylglycerol (DG) to phosphatidic acid. The enzyme contains EF-hand motifs, cysteine-rich zinc-finger-like sequences and putative ATP-binding sites conserved in other kinase species. This enzyme is found mainly in neurons, localized in their cell membranes.[82]

7. EF-HAND CA²⁺-BINDING PROTEINS IN THE ENDOPLASMIC RETICULUM

A major Ca^{2+} storage compartment in eukaryotic cells is the endoplasmic reticulum (ER) in non-muscle cells or the sarcoplasmic reticulum (SR) in muscle cells. Major soluble proteins in the human ER include, for example, endoplasmin and calreticulin, the latter being the major Ca^{2+} storage protein. None of these proteins are members of the EF-hand family and, instead, coordinate Ca^{2+} via a cluster of acidic amino acids.[80] Recently two novel resident ER proteins were identified that constitute a new subfamily of the EF-hand superfamily of Ca^{2+}-binding proteins. The protein termed endoplasmic reticulum Ca^{2+}-binding protein of 55 kD (ERC-55) contains six copies of the EF-hand motif.[81] The highest sequence identity was found to reticulocalbin, also containing six EF-hand motifs. The biological function of these proteins is not known but it is proposed that they might regulate Ca^{2+}-dependent activities in the ER by modulating the functions of other proteins.

8. S100 PROTEINS

S100 proteins are low molecular weight proteins of 9-12 kD. Originally isolated from brain, a large variety of tissues have now been shown to express one or more members of the S100 family.[83,84] At present, 13 different S100 proteins have been isolated and characterized from human tissues and a new nomenclature has been proposed[85] (Table 2.1). S100 proteins are involved in many different cellular activities such as cell cycle progression and differentiation[83,84] and in tumor progression,[86-88] and extracellular functions, e.g. S100β-dimer in promoting neurite extension activity.[89]

Two members of the S100 protein family have been crystallized, calbindin 9k[90] and S100β.[91] All S100 proteins have a common domain structure with two hydrophobic regions at the

N- and C-termini. The N-terminal EF-hand (containing 14 amino acids) is located in a region rich in basic amino acids; the C-terminal EF-hand (containing 12 amino acids) is located in a region rich in acidic amino acids. The region of maximum divergence of S100 proteins is in the 'hinge' region between the two EF-hands. Ca^{2+} binds first to the high affinity C-terminal EF-hand. This induces a conformational change in the C-terminal hydrophobic domain and exposure of this domain to the solvent. At higher concentrations Ca^{2+} is bound to the N-terminal hydrophobic domain resulting in exposure of this domain to the solvent. These different conformations are assumed to interact with different binding proteins. The Ca^{2+}-binding affinity of the C-terminal EF-hand is K_D=20-50 μM and the affinity for the N-terminal EF-hand is K_D=200-500 μM.[83,84]

Similar values have been reported for the two human S100 proteins, $S100A_6$ (calcyclin) and $S100A_4$ (CAPL).[92] In addition, it has been shown that these sites are insensitive to Mg^{2+} and are therefore of the Ca^{2+}-specific type.

In addition to Ca^{2+}, $S100A_6$ also binds Zn^{2+},[93] as does $S100\beta$[94] or the newly discovered calgranulin.[95] Binding of metal ions other than Ca^{2+} is also suggested for the newly discovered $S100A_3$ (S100E) which exhibits unique features within the S100 protein family,[96] having the highest content of cystein (10 residues) of all calcium-binding proteins. Post translational modification (phosphorylation) of $S100A_9$ (p14)[97] has also been reported. The level of phosphorylation was found to be related to the increase of intracellular Ca^{2+}, an additional mode of control.

S100 proteins interact with a number of target proteins (enzymes, components of the cytoskeleton and annexins). It has been suggested that an increase of intracellular Ca^{2+} and binding to S100 proteins results in a conformational change, exposure of hydrophobic domains and interaction with target proteins.

When intracellular Ca^{2+} concentration returns to basal levels the S100 proteins may dissociate from their binding protein. The monomeric/dimeric form might bind to different target proteins. It is also assumed that the dimeric form may be secreted from cells (chapter 4).

The mechanism of secretion is not known since S100 proteins do not possess a classical leader peptide guiding the protein to the secretory pathway. However, one possible mechanism might be an interaction of the hydrophobic domain (exposed when Ca^{2+} is bound) of the S100 protein with a membrane protein (e.g. annexin) leading to secretion.

Future goals are to identify the tissue- and cell-specific expression of S100 proteins, their route of secretion, and their intra- and extracellular target proteins (especially annexin), in the hope that this may provide clues to their functions.

Some novel Ca^{2+}-binding proteins have recently been detected in the brain, including R_2D_2 antigen,[98] calciphosine[99] homologous to CCBP-23,[100] but their exact localization and functions have not yet been examined.

B. ANNEXINS, CALCIUM- AND PHOSPHOLIPID-BINDING PROTEINS.

The annexin family presently consists of 13 members expressed in a wide variety of organisms (selected reviews[101-109]). Several names have been ascribed to members of this family, such as lipocortins, calelectrins, calcimedins, calpactins, chromobindins and endonexins. A common nomenclature with the generic term, annexin, has been introduced (Table 2.3).

Based on their sequence homologies, annexins bind to acidic phospholipids in a Ca^{2+}-dependent manner but lack the 'EF-hand' Ca^{2+}-binding motif which is characteristic for calmodulin, parvalbumin, S100 proteins and many other members of that large protein family.

Annexins are characterized by a conserved 70-amino acid domain repeated either four or eight times in the protein structure and the ability to bind to phospholipids in a Ca^{2+}-dependent manner (Fig. 2.1, bottom). The conserved domains contain the calcium- and phospholipid-binding sites.

The sequences of the N-terminal domains of the annexins show few sequence similarities, suggesting functional individuality. The N-termini of annexin I and II contain sites for phosphorylation by tyrosine and serine/threonine kinases. Annexin II binds $S100A_{10}$ (p11,[110]) and annexins II, VI, and XI bind $S100A_6$ (calcyclin).[111,112]

Table 2.4 lists the chromosomal assignments of the genes encoding human annexins. Table 2.5 summarizes the localization of annexins, their proposed physiological roles and disorders where an abnormal expression of annexins have been found. Annexins are reported to be anti-inflammatory, to inhibit blood coagulation, to be involved in exocytosis, cytoskeletal organization, cell adhesion and membrane trafficking. Considerable progress has been made by the analysis of the three-dimensional crystal structure of annexin V. Based on these studies it has been proposed that annexin V can function as a Ca^{2+} channel (Table 2.5).

A cell-specific expression of annexins has also been reported in the central nervous system, suggesting a unique function for annexin V in glial cells and for annexin VI in neuronal cells.

Recently it has been proposed that annexin I has a protective effect against ischemic brain damage via modulation of the release of arachidonic acid, influencing neuronal death. This could raise possibilities of novel approaches to therapeutic interventions.[113,114]

Annexins have also been reported to be increased in the central nervous system in multiple sclerosis and these findings are discussed in relation to the inflammatory processes in the CNS and steroid therapy of multiple sclerosis.

Table 2.4. Chromosomal assignment of the genes encoding human annexins

Annexin	Chromosome	Selected Refs
I	9q11-q22	1
II	15q21-q22	1,2
III	4q21	3
IV	2p13	4
V	4q26-q28	5
VI	5q32-q34	6
VII	10q21-q23	7
VIII	10q	8

The chromosomal loci of the human annexin genes are dispersed throughout the genome and do not form a linked gene cluster. Interestingly, annexin VI was mapped within 2.1 cM of the spasmody locus. Spasmody is a recessive mouse mutation characterized by a prolonged righting reflex, fine motor tremor, leg clasping, and stiffness. Syntenic homology between human chromosome 5q and mouse chromosome 11 and phenotypic similarities suggest that spasmody mice may be a genetic model for the inherited human startle disease, hyperexplexia.[9]

---

1. Huebner K, Cannizzaro LA, Frey AZ et al. Oncogene Res 1988; 2:299-310.
2. Spano F, Raugei G, Palla E et al. Gene 1990; 95: 243-251.
3. Tait JF, Smith C, Yu L et al. Genomics 1993; 18:79-86.
4. Tait JF, Smith C, Frankenberry DA et al. Genomics 1992; 12:313-318.
5. Tait JF, Frankenberry DA, Shiang R et al. Cytogenet Cell Genet 1991; 57:187-192.
6. Davies A, Moss SE, Crompton MR et al. Hum Genet 1989; 82:234-238.
7. Burns AL, Shirvan A, McBride OW et al. Am J Hum Genet 1991; 49:401 (abstract).
8. Chang KS, Wang G, Freireich EJ et al. Blood 1992; 79:1802-1810.
9. Buckwalter MS, Testa CM, Noebels JL et al. Genomics 1993; 17:279-286.

Table 2.5. The annexins, Ca^{2+} and phospholipid-binding proteins

Annexins	Expression/localization	Suggested intra- and extracellular functions	Association with diseases
I	glial cells and neurons[1-5]; skin (keratinocytes)[6]; may extend from cytoplasmic face of cell membranes to extracellular space[7]	cytosolic and secreted protein; crystal structure is known[8]; N-terminus contains phosphorylation sites for epidermal growth factor; receptor and protein kinase C; cell transformation; stimulation of cell growth; regulation of cytoskeletal organization; exocytosis via membrane - membrane fusion and mediation of the anti-inflammatory action of steroids (via inhibition of phospholipase A_2); Ca^{2+} channel activity[7,9,10]	chronic rheumatic diseases (systemic lupus erythematosus and rheumatoid arthritis),[10-13] inflammatory CNS diseases; multiple sclerosis[14]; Mediation of impaired febrile responses[15]; ischemic brain damage[16]; altered expression (also of other annexins) in pathological conditions of the brain[17]
II	cytoplasmic face of membrane; many cell lines and tissues; most abundant in intestinal epithelium, lung and placenta;[18-20] chromaffin cells of the adrenal medulla;[21] highly expressed in glioblastoma and astrocytoma[22]	complex with the S100 A_{10} (p11) (annexin II$_2$ p11$_2$); substrate for pp60^{src} and other kinases;[23] cellular growth and differentiation; chromaffin granule fusion and inhibition of blood coagulation[24-26]; mediator of exocytosis in some cells;[21] binding to cytomegalovirus[27]; binds to DNA[28]; involved in chromosome DNA replication and cell proliferation[29,30]; high affinity receptor for tenascin-C mediating cellular responses to the extracellular matrix;[31] a novel N-terminal variant of annexin VI was found in leukemia cells[32]	anti-inflammatory[9,10,33] pathogenesis of glioblastoma[22]; inhibition of blood coagulation[33]; inflammatory CNS diseases and increased in patients with multiple sclerosis[34]
III	spleen, reproductive and other tissues[35]	specialized function, restricted expression[35]	anti-inflammatory inhibition of blood coagulation

(Continued)

(Table 2.5 continued)

	Tissue distribution	Properties / function	Clinical / disease association
IV	reproductive and other tissues[35]	X-ray structure known;[36] inhibition of calcium-activated chloride conductance[37]	anti-inflammatory inhibition of blood coagulation; increased in patients with multiple sclerosis[34]
V	high levels expressed in several tissues, including brain, heart, lung, liver, and reproductive tissues[35,38,39,40]	X-ray structure known[38,39,41-48]	anti-inflammatory inhibition of blood coagulation;
	localized in the cytoplasm and associated with intracellular vesicles and plasma membranes, also found extracellularly and in body fluids	Ca^{2+} channel activity[49]	inflammatory CNS diseases and increased in patients with multiple sclerosis[34]
		binds to actin; anti-coagulant activity and antiphospholipase activities; neurotrophic factor in the CNS[50]	
VI	staining of Purkinje cells and of other neuronal elements in the cerebellum;[51] high concentrations in muscle, heart, lung, pancreas, and neutrophils,[35,38,39,52] binding to intracellular vesicles	8 instead of 4 tandem repeats; X-ray structure[53] and gene structure known[54] modulates the release channel activity of the sarcoplasmic reticulum[55] inhibitor of phospholipase A$_2$; inhibitor of annexin II-mediated chromaffine granule fusion; essential for budding of clathrin-coated pits[56] but this has been questioned;[57] interaction with a membrane skeletal protein, calspectin, in brain,[58] cell growth regulation[57]	anti-inflammatory inhibition of blood coagulation; cell growth regulation[57]
VII	several mammalian tissues, including liver, brain, skeletal and cardiac muscle[59]	may function as a Ca^{2+} channel[55,60-62] chromaffine granule aggregation and exocytosis[63]	
VIII	expressed in lung, skin, liver, kidney;[40] high levels in acute myelocytic leukemia[64]	anti-coagulant activity; anti-phospholipase activity[65] specialized function in lung endothelium[40]	associated with the particular abnormal hemostasis of some leukemias[64]

IX and X	arthropods (Drosophila)[66]	
XI	widely distributed, mainly in nuclei of fibroblasts[67]	binds S100A₆ (calcyclin); autoimmune diseases[69]
		is phosphorylated[67,68]
XII	identified in hydra vulgaris[70]	phosphorylated by mitogen-activated protein kinase[71]
XIII	intestine specific annexin[72]	

1. Johnson MD, Kamso-Pratt J, Pepinsky RB et al. Am J Clin Pathol 1989; 92:424-430.
2. Johnson MD, Kamso-Pratt J, Pepinsky RB et al. Hum Pathol 1989; 20:772-776.
3. Smillie F, Bolton C, Peers SH et al. Br J Pharmacol 1989; 97:90-95.
4. Strijbos PJLM, Tilders FJH, Carey F et al. Brain Res 1990; 553:249-260.
5. McKanna JA. (1993) J. Neurosci. Res. 36:491-500.
6. Fava RA, Nanney LB, Wilson D et al. J Invest Dermatol 1993; 101:732-737.
7. Haigler HT, Schlaepfer DD. In: Moss SE, ed. The Annexins. London - Chapel Hill: Portland Press, 1992:11-22.
8. Weng X, Luecke H, Song S et al. Protein Sci 1993; 2:448-458.
9. Russo-Marie F. In: Moss SE, ed. The Annexins. London - Chapel Hill: Portland Press, 1992:35-46.
10. Cirino G, Flower RJ. In: Heizmann CW, ed. Novel Calcium-Binding Proteins. Fundamentals and Clinical Implications. Berlin, Springer-Verlag. 1991:589-611.
11. Hirata F, Del Carmine R, Nelson CA. Proc Natl Acad Sci USA 1981; 78:3190-3194.
12. Goulding HJ, Podgorski MR, Hall ND et al. Ann Rheum Diseases 1989; 48:843-850.
13. Browning JL, Ward MP, Wallner BP et al. In: Cytokines and Lipocortin in Inflammation and Differentiation. New York: Wiley-Liss Inc., 1990:27-45.
14. Bolton C, Flower RJ. J Neuroimmun 1992; 39:91-100.
15. Strijbos PJLM, Horan MA, Carey F et al. Am J Physiol 1993; 265:E289-E297.
16. Rothwell NJ, Relton JK. Cerebrovasc Brain Metab Rev 1993; 5:178-198.
17. Eberhard DA, Brown MD, VanderBerg SR. Am J. Pathol 1994; 145:640-649.
18. Zokas L, Glenney JR Jr. J Cell Biol 1987; 105:2111-2121.
19. Gould KL, Cooper JA, Hunter T. J Cell Biol 1984; 98:487-497.
20. Saris CJM, Kristensen T, D'Eustachio P et al. J Biol Chem 1987; 262:10663-10671.
21. Burgoyne RD. In: Moss SE, ed. The Annexins. London - Chapel Hill: Portland Press, 1992:69-76.
22. Reeves SA, Chavez-Kappel C, Davis R et al. Cancer Res 1992; 52:6871-6876.
23. Gerke V. In: Heizmann CW, ed. Novel Calcium-Binding Proteins. Fundamentals and Clinical Implications. Berlin: Springer-Verlag 1991; 139-155.
24. Tait JF, Sakata M, McMullen BA et al. Biochemistry 1988; 27:6268-6276.
25. Ali SM, Geisow MJ, Burgoyne RD. Nature (London) 1989; 340:313-315.
26. Drust DS, Creutz CE. Nature (London) 1988; 331:88-91.
27. Wright JF, Kurosky A, Wasi S. Biochem Biophys Res Commun 1994; 198:983-989.
28. Boyko V, Mudrak O, Svetlova M et al. FEBS Lett 1994; 345:139-142.
29. Vishwanatha JK, Kurosky A, Wasi S. J Cell Sci 1993; 105:533-540.
30. Vishwanatha JK, Chiang Y, Kumble DK et al. Carcinogenesis 1993; 14:2575-2579.
31. Chung CY, Erickson HP. J Cell Biol 1994; 126:539-548.
32. Upton AL, Moss SE. Biochem J 1994; 302:425-428.
33. Russo-Marie F. In: Smith VL, Dedman JR, eds. Stimulus Response Coupling: The Role of Intracellular Calcium-binding Proteins. Boca Raton, Florida: CRC Press, 1990:467-481.
34. Elderfield A-J, Newcombe J, Bolton C et al. J Neuroimmun 1992; 39:91-100.
35. Yeatman TJ, Updyke TV Kaetzel MA et al. Clin. Exp. Metastasis 1993; 11:37-44.
36. Boustead CM, Walker JH, Kennedy D et al. FEBS Lett 1991; 279:187-189.
37. Kaetzel MA, Chan HC, Dubinsky WP et al. J Biol Chem 1994; 269:5297-5302.
38. Giambanco T, Verzini M, Donato R. Biochem Biophys Res Commun 1993; 196:1221-1226.
39. Bewley MC, Boustead CM, Walker JH et al. Biochemistry 1993; 32:3923-3929.
40. Reutlingsperger CPM, van Heerde W, Hauptmann R et al. FEBS Lett 1994; 349:120-124.
41. Rojas E, Pollard HB, Haigler H et al. J Biol Chem 1990; 265:21207-21215.
42. Lewit-Bentley A, Doublie S, Fourme R et al. J Mol Biol 1989; 210:875-876.
43. Seaton BA, Head JF, Kaetzel MA et al. J Biol Chem 1990; 265:4567-4569.
44. Huber R, Römisch J, Paques E. EMBO J 1990; 9:3867-3874.
45. Huber R, Schneider M, Mayr I et al. FEBS Lett 1990; 275:15-21.
46. Mosser G, Ravanat C, Freyssinet J-M et al. J Mol Biol 1991; 217:241-245.
47. Brisson A, Mosser G, Huber R. J Mol Biol 1991; 220:199-203.
48. Concha NO, Head JF, Kaetzel MA et al. Science 1993; 261:1321-1324.
49. Burger A, Voges D, Demange P et al. J Mol Biol 1994; 237:479-499.
50. Takei N, Ohsawa K, Imai Y et al. Neurosci Lett 1994; 171:59-62.
51. Boustead CM, Kenny AJ, Vaughan FT et al. Biochem Soc Transact 1993; 21:292S (Abstract).
52. Doubell AF, Lazure C, Charbonneau C et al. Cardiovasc Res 1993; 27:1359-1367.
53. Newman R, Tucker A, Ferguson C et al. J Mol Biol 1989; 206:213-219.
54. Smith PD, Davies A, Crumpton MJ. Proc Natl Acad Sci USA 1994; 91:2713-1717.
55. Pollard HB, Burns AL, Rojas E. J Membr Biol 1990; 117:101-112.
56. Lin HC, Südhof TC, Anderson RGW et al. Cell 1992; 70:283-291.
57. Smythe E, Smith PD, Jacob SM et al. J Cell Biol 1994; 124:301-306.
58. Watanabe T, Inui M, Chen B-Y et al. J Biol Chem 1994; 269:17656-17662.
59. Magendzo K, Shirvan A, Cultraro C et al. J Biol Chem 1991; 266:3228-3232.
60. Burns AL, Magendzo K, Shirvan A et al. Proc Natl Acad Sci USA 1989; 86:3798-3802.
61. Pollard HB, Rojas E. Proc Natl Acad Sci USA 1988; 85:2974-2978.
62. Pollard HB, Guy HR, Arispe N et al. Biophys J 1992; 62:19-22.
63. Pollard HB, Rojas E, Merezhinskaya N et al. In: Moss SE, ed. The Annexins. London - Chapel Hill: Portland Press, 1992:89-103.
64. Hu Z-B, Ma W, Uphoff CC et al. Leukemia Res 1993; 17:949-957.
65. Hauptmann R, Maurer-Fogy I, Krystek E et al. Eur J Biochem 1989; 185:63-71.
66. Smith PD, Moss SE. Trends Genetics 1994; 10:241-246.
67. Mizutani A, Tokumitsu H, Kobayashi R et al. J Biol Chem 1993; 268: 15517-15572.
68. Towle CA, Treadwell BV. J Biol Chem 1992; 267:5416-5423.
69. Misaki Y, Pruijn GJM, van der Kemp AWCM, et al. J Biol Chem 1994; 269: 4240-4246.
70. Schlaepfer DD, Fisher DA, Brandt ME et al. J Biol Chem 1992; 267:9529-9539.
71. Gupta SK, Gallego C, Johnson GL et al. J Biol Chem 1992; 267:7987-7990.
72. Wice BM, Gordon JI. J Cell Biol 1992; 116:405-422.

REFERENCES

1. Kretsinger RH. Structure and evolution of calcium-modulated proteins. CRC Crit Rev Biochem 1980; 8:119-174.

2. Heizmann CW, Hunziker W. Intracellular calcium-binding proteins: more sites than insights. Trends Biochem Sci 1991;
   16:98-103.

3. Kretsinger RH, Nockolds CW. Carp muscle calcium-binding protein. II. Structure determination and general description. J Biol
   Chem 1973; 248:3313-3326.

4. Cohen P, Klee CB, eds. Calmodulin. In: Molecular Aspects of Cellular Regulation. Vol. 5. New York: Elsevier, 1988

5. Nojima H. Structural organization of multiple rat calmodulin genes.
   J Mol Biol 1989; 208:269-282.

6. DeLorenzo RJ. Calmodulin in neurotransmitter release and synaptic function. Fed Proc 1982; 41:2265-2272.

7. Ikeshima H, Yuasa S, Matsuo K et al. Expression of three nonal-
   lelic genes coding calmodulin exhibits similar localization on the
   central nervous system of adult rats. J Neurosci Res 1993;
   36:111-119.

8. Weinman J, Gaspera BD, Dautigny A et al. Developmental regulation of calmodulin gene expression in rat brain and skeletal muscle.
   Cell Regul 1991; 2:819-826.

9. Heizmann CW. Parvalbumin, an intracellular Ca^{2+}-binding protein;
   distribution, properties and possible roles in mammalian cells.
   Experientia (Basel) 1984; 40:910-921.

10. Goodman M, Pechère JF. The evolution of muscular parvalbumins
    investigated by the maximum parsimony method. J Mol Evol 1977;
    9:131-158.

11. Baldellon C, Padilla A, Care A. Kinetics of amide proton exchange
    in parvalbumin studied by ^{1}H 2-D NMR. A comparison of the
    calcium and magnesium loaded forms. Biochimie (Paris) 1992;
    74:837-844.

12. Roquet F, Declerq J-P, Tinant B et al. Crystal structure of the
    unique parvalbumin component from muscle of the leopard shark
    (*Triakis semifasciata*). The first X-ray study of an α-parvalbumin. J
    Mol Biol 1992; 223:705-720.

13. Declerq J-P, Tinant B, Parello J et al. Ionic interactions with
    parvalbumins. Crystal structure determination of pike 4.10
    parvalbumin in four different ionic environments. J Mol Biol 1991;
    220:1017-1039.

14. Ahmed FR, Przybylska M, Rose DR et al. Structure of oncomodulin
    refined at 1.85 Å resolution. J Mol Biol 1990; 216:127-140.

15. Kuster T, Staudenmann W, Hughes GJ et al. Parvalbumin isoforms
    in chicken muscle and thymus. Amino acid sequence analysis of
    muscle parvalbumin by tandem mass spectrometry. Biochemistry.
    1991; 30:8812-8816.

16. Schleef M, Zühlke C, Jockusch H et al. The structure of the mouse parvalbumin gene. Mamm Genom 1992; 3:217-225.

17. Brewer JM, Arnold J, Beach GG et al. Comparison of the amino acid sequences of tissue-specific parvalbumins from chicken muscle and thymus and possible evolutionary significance. Biochem Biophys Res Commun 1991; 181:226-231.

18. Hall CA, Beach GG, Ragland WL. Monoclonal antibody for avian thymic hormone. Hybridoma 1991; 10:575-582.

19. Föhr UG, Weber BR, Müntener M et al. Human α and β parvalbumins. Structure and tissue-specific expression. Eur J Biochem 1993; 215:719-727.

20. Hauer CR, Staudenmann W, Kuster T et al. Protein sequence determination by ESI-MS and LSI-MS tandem mass spectrometry: parvalbumin primary structures from cat, gerbil and monkey skeletal muscle. Biochim Biophys Acta 1992; 1160:1-7.

21. Brewer LM, MacManus JP. Detection of oncomodulin, an onodevelopmental protein in human placenta and choriocarcinoma cell lines. Placenta 1987; 8:351-363.

22. Huber S, Leuthold M, Sommer EW et al. Human tumor cell lines express low levels of oncomodulin. Biochem Biophys Res Commun 1990; 169:905-909.

23. Christakos S, Gabrielides C, Rhoten WB. Vitamin D-dependent calcium-binding proteins. Chemistry, distribution, functional considerations and molecular biology. Endocr Rev 1989; 10:3-26.

24. Feher JJ. Measurement of facilitated calcium diffusion by a soluble calcium-binding protein. Biochim Biophys Acta 1984; 773:91-98.

25. Wilson PW, Harding M, Lawson DEM. Putative amino acid sequence of chick calcium-binding protein deduced from a complementary DNA sequence. Nucleic Acids Res 1985; 13:8867-8881.

26. Hunziker W. The 28-kDa vitamin D-dependent calcium-binding protein has a six-domain structure. Proc Natl Acad Sci USA 1986; 83:7578-7582.

27. Yamakuni T, Kuwano R, Odani S et al. Nucleotide sequence of cDNA to mRNA for cerebellar Ca²⁺-binding protein, spot 35 protein. Nucleic Acids Res 1986; 14:6768.

28. Parmentier M, Lawson DEM, Vassant G. Human 27 kDa calbindin complementary DNA sequence. Evolutionary and functional implication. Eur J Biochem 1987; 170:207-215.

29. Hunziker W, Schrickel S. Rat brain calbindin D-28K: Six domain structure and extensive amino acid sequence homology with chick calbindin D-28K. Mol Endocrinol 1988; 2:465-473.

30. Kumar R, Litwiller RD, Gross M et al. A comparison of the amino terminal sequence of rat and human vitamin D-binding proteins. Fed Proc 1986; 45:826 (Abstract).

31. Takagi T, Konishi K, Cox JA. Amino acid sequence of two sarcoplasmic calcium binding proteins from Protochordate Amphioxus. Biochemistry 1986a; 25:3585-3592.

32. Fullmer CS, Wasserman RH. Chicken intestinal 28-kilodalton calbindin-D: Complete amino acid sequence and structural considerations. Proc Natl Acad Sci USA 1987; 84:4772-4776.

33. Hunziker W. The 28-kDa vitamin D-dependent calcium-binding protein has a six-domain structure. Proc Natl Acad Sci USA 1986; 83:7578-7582.

34. Bredderman PJ, Wasserman RH. Chemical composition, affinity for calcium and some related properties of the vitamin D-dependent calcium-binding proteins. Biochemistry 1974; 14:1687-1694.

35. Rogers JH. Calretinin. In: Heizmann CW, ed. Novel Calcium-Binding Proteins. Berlin: Springer-Verlag, 1991:251-276.

36. Winsky L, Jacobowitz DM. In: Heizmann CW, ed. Novel Calcium-Binding Proteins. Fundamentals and Clinical Implications. Berlin: Springer-Verlag, 1991:277-300.

37. Ellis J, Rogers J. Design and specificity of hammerhead ribozymes against calretinin mRNA. Nucleic Acids Res 1993; 21:5171-5178.

38. Palcziewski K, Subbaraya I, Gorczyca WA, et al. Molecular cloning and characterization of retinal photoreceptor quanylyl cyclase-activating protein. Neuron 1994; 13:395-404.

39. Kawamura S, Takamatsu K, Kitamura K. Purification and characterization of S-modulin, a calcium-dependent regulator on cGMP phosphodiesterase in frog rod photoreceptors. Biochem Biophys Res Commun 1992; 186:411-417.

40. Yamagata K, Goto K, Kuo C-H et al. Visinin: a novel calcium binding protein expressed in retinal cone cells. Neuron 1990; 2:469-476.

41. Kobayashi M, Takamatsu K, Saitoh S et al. Molecular cloning of hippocalcin, a novel calcium-binding protein of the recoverin family exclusively expressed in hippocampus. Biochem Biophys Res Commun 1992; 189:511-517.

42. Kuno T, Kajimoto Y, Hashimoto T et al. cDNA cloning of a neural visinin-like Ca^{2+}-binding protein. Biochem Biophys Res Commun 1992; 184:1219-1225.

43. Lenz SE, Henschel Y, Zopf D et al. VILIP, a cognate protein of the retinal calcium binding proteins visinin and recoverin, is expressed in the developing chicken brain. Mol Brain Res 1992; 15:133-140.

44. Okazaki K, Watanabe M, Ando Y et al. Full sequence of neurocalcin, a novel calcium-binding protein abundant in central nervous system. Biochem Biophys Res Commun 1992; 185: 147-153.

45. Takamatsu K, Kitamura K, Noguchi T. Isolation and characterization of recoverin-like Ca^{2+}-binding protein from rat brain. Biochem Biophys Res Commun 1992; 183:245-251.

46. Terasawa M, Nakano A, Kobayashi R et al. Neurocalcin: a novel calcium-binding protein from bovine brain. J Biol Chem 1992; 267:19596-19599.

47. Lipp H-P, Wolfer DP, Qin WX et al. CBP-18, a Ca²⁺-binding protein in rat brain: tissue distribution and localization. J Neurochem 1993; 60:1639-1649; see appendix J Neurochem 1993; 61:790.
48. Pongs O, Lindemeier J, Zhu XR et al. Frequenin - a novel calcium-binding protein that modulates synaptic efficacy in the Drosophila nervous system. Neuron 1993; 11:15-28.
49. Flaherty KM, Zozulya S, Stryer L et al. Three-dimensional structure of recoverin, a calcium sensor in vision. Cell 1993; 75:709-716.
50. Murakami A, Yajima T, Inana G. Isolation of human retinal genes: recoverin cDNA and gene. Biochem Biophys Res Commun 1992; 187:234-244.
51. Wiechmann AF, Akots G, Hammarback JA et al. Genetic and physical mapping of human recoverin: a gene expressed in retinal photoreceptors. Investigative Ophthalm & Visual Sci 1994; 35:325-331.
52. Milam AH, Dacey D, Dizhoor AM. Recoverin immunoreactivity in mammalian cone bipolar cells. Vis Neurosci 1993; 10:1-2.
53. Polans AS, Burton MD, Haley TL et al. Recoverin, but not visinin, is an autoantigen in the human retina identified with a cancer-associated retinopathy. Investigative Ophthalm & Visual Sci 1993; 34:81-90.
54. Adamus G, Guy J, Schmied JL et al. Role of anti-recoverin autoantibodies in cancer-associated retinopathy. Investigative Ophthalmol & Visual Sci 1993; 34:2626-2633.
55. Thirkill CE, Tait RC, Tyler NK et al. The cancer-associated retinopathy antigen is a recoverin-like protein. Investigative Ophthalmol & Visual Sci 1992; 33:2768-2772.
56. Hurley JB, Dizhoor AM, Ray S et al. Recoverin's role: conclusion withdrawn. Science 1993; 260:740.
57. Klee CB, Cohen P. The calmodulin-regulated protein phosphatase in molecular aspects of cellular regulation. In: Cohen P, Klee CB, eds. Calmodulin. Vol. 5. Amsterdam: Elsevier, 1988:225-247.
58. Klee CB, Draetta GF, Hubbard MJ. Calcineurin. Adv Enzymol 1988; 61:149-200.
59. Kuno T, Mukai H, Ito A et al. Distinct cellular expression of calcineurin Aα and Aβ in rat brain. J Neurochem 1992; 58:1643-1651.
60. Guerini D, Klee CB. Cloning of human calcineurin A: Evidence for two isozymes and identification of a polyproline structural domain. Proc Natl Acad Sci USA 1989; 86:9183-9187.
61. Kincaid RL, Rathna Giri P, Higuchi S et al. Cloning and characterization of molecular isoforms of the catalytic subunit of calcineurin using nonisotopic methods. J Biol Chem 1990; 265:11312-11319.
62. Muramatsu T, Kincaid RL. Molecular cloning and chromosomal mapping of the human gene for the testis-specific catalytic subunit of calmodulin-dependent protein phosphatase (calcineurin A). Biochem Biophys Res Commun 1992; 188:265-271.

63. Liu J, Farmer JD Jr, Lane WS et al. Calcineurin is a common target of cyclophilin-cyclosporin A and FKBP.FK506 complexes. Cell 1991; 66:807-815.

64. Steiner JP, Dawson TM, Fotuhi M et al. High brain densities of the immunophilin FKBP colocalized with calcineurin. Nature 1992; 358:584-585.

65. Polli JW, Billingsley ML, Kincaid RL. Expression of the calmodulin-dependent protein phosphatase, calcineurin, in rat brain: developmental patterns and the role of nigrostriatal innervation. Devel Brain Res 1991; 63:105-119.

66. Liu J-P, Sim ATR, Robinson PJ. Calcineurin inhibition of dynamin I GTPase activity coupled to nerve terminal depolarization. Science 1994; 265:970-973.

67. Mulkey RM, Endo S, Shenolikar S et al. Involvement of a calcineurin/inhibitor-1 phosphatase cascade in hippocampal long-term depression. Nature 1994; 369:486-487.

68. Ohno S, Emori Y, Imajoh S et a. Evolutionary origin of a calcium-dependent protease by fusion of genes for a thiol protease and a calcium-binding protein? Nature 1984; 312:566-70.

69. Sakihama T, Kakidani H, Zenita K et al. Proc Natl Acad Sci USA 1985; 82:6075-6079.

70. Emori Y, Kawasaki H, Sugihara H et al. Isolation and sequence analyses of cDNA clones for the large subunits of two isozymes of rabbit calcium-dependent protease. J Biol Chem 1986; 261: 9465-9471.

71. Suzuki K. Calcium activated neutral protease: domain structure and activity regulation. TIBS 1987; 12:103-105.

72. Murachi T. Intracellular regulatory system involving calpain and calpastatin. Biochem Intern 1989; 18:263-294.

73. Croall DE, Demartino GN. Calcium-activated neutral protease (calpain) system: structure, function, and regulation. Physiol Rev 1991; 71:813-847.

74. Saido TC, Sorimachi H, Suzuki K. Calpain: new perspectives in molecular diversity and physiological-pathological involvement. FASEB J 1994; 8:814-822.

75. Hamakubo T, Kannagi R, Murachi T et al. Distribution of calpains I and II in rat brain. J Neurosci 1986; 6:3103-3111.

76. Shimohama S, Suenaga T, Araki W et al. Presence of calpain II immunoreactivity in senile plaques in Alzheimer's disease. Brain Res 1991; 558:105-108.

77. Iwamoto N, Thangnipon W, Crawford C et al. Localization of calpain immunoreactivity in senile plaques and in neurons undergoing neurofibrillary degeneration in Alzheimer's disease. Brain Res 1991; 561:177-180.

78. Saito K-I, Elce JS, Hamos JE et al. Widespread activation of calcium-activated neutral proteinase (calpain) in the brain in Alzheimer

disease: A potential molecular basis for neuronal degeneration. Proc Natl Acad Sci USA 1993; 90:2628-2632.

79. Sorimachi H, Saido TC, Suzuki K. New area of calpain research; discovery of tissue-specific calpains. FEBS Lett 1994; 343:1-5.

80. Sambrook JF. The involvement of calcium in transport of secretory proteins from the endoplasmic reticulum. Cell 1990; 61:197-199.

81. Weis K, Griffiths G, Lamond AI. The endoplasmic reticulum calcium-binding protein of 55 kDa is a novel EF-hand protein retained in the endoplasmic reticulum by a carboxy-terminal His-Asp-Glu-Leu motif. J Biol Chem 1994; 269:19142-19150.

82. Goto K, Kondo H. Molecular cloning and expression of a 90-kDa diacylglycerol kinase that predominantly localizes in neurons. Proc Natl Acad Sci USA 1993; 90:7598-7602.

83. Kligman D, Hilt DC. The S100 protein family. Trends Biochem Sci 1988; 13:437-443.

84. Hilt DC, Kligman D. The S-100 protein family: A biochemical and functional overview. In: Heizmann CW, ed. Novel Calcium-binding Proteins. Berlin: Springer Verlag, 1991:65-103.

85. Schäfer BW, Wicki R, Engelkamp D. Isolation of a YAC clone covering a cluster of nine S100 genes on human chromosome 1q21: Rationale for a new nomenclature of the S100 calcium-binding protein family. Genomics 1995; 25:638-643.

86. Lee SC, Kim IG, Marekov LN et al. The structure of human trichohyalin: potential multiple functions as an EF-hand-like calcium binding protein, cornified cell envelope precursor and an intermediate filament associated (crosslinking) protein. J Biol Chem 1993; 268:12164-12176.

87. Davies BR, Davies M, Gibbs F et al. Induction of the metastatic phenotype by transfection of a benign rat mammary epithelial cell line with the gene for p9Ka, a rat calcium-binding protein, but not with the oncogene EJ-ras-1. Oncogene 1993; 8:999-1008.

88. Pedrocchi M, Schäfer BW, Mueller H et al. Expression of Ca²⁺-binding proteins of the S100 family in malignant human breast-cancer cell lines and biopsy samples. Int J Cancer 1994; 57:684-690.

89. Selinfreund RH, Barger SW, Pledger WJ et al. Neurotrophic protein S100β stimulates glial cell proliferation. Proc Natl Acad Sci USA 1991; 88:3554-3558.

90. Szebenyi DME, Obendorf SK, Moffat K. Structure of vitamin D-dependent calcium-binding protein from bovine intestine. Nature (London) 1981; 294:327-332.

91. Charles RS, Kumar VD. Crystallization and preliminary X-ray analysis of apo-S100β and S100β with Ca²⁺. J Mol Biol 1994; 236:953-957.

92. Pedrocchi M, Schäfer BW, Durussel I et al. Purification and characterization of the recombinant human calcium-binding S100 proteins CAPL and CACY. Biochemistry 1994; 33:6732-6738.

93. Filipek A, Heizmann CW, Kuznicki J. Calcyclin is a calcium- and zinc-binding protein. FEBS-Lett 1990; 264:263-266.
94. Baudier J, Glasser N, Gerrard D. Ion binding to S100 proteins. I. Calcium- and zinc-binding properties to bovine brain S100$\alpha\alpha$, S100a(α,β) and S100b($\beta\beta$): Zn^{2+} regulates Ca^{2+}-binding on S100β protein. J. Biol Chem. 1986; 261:8192-8203.
95. Dell'Angelica EC, Schleicher CH, Santomé JA. Primary structure and binding properties of calgranulin C, a novel S100-like calcium-binding protein from pig granulocytes. J Biol Chem 1994; 269:28929-28936.
96. Engelkamp D, Schäfer BW, Mattei MG et al. Six S100 genes are clustered on human chromosome 1q21: Identification of two genes coding for the two previously unreported calcium-binding proteins S100D and S100E. Proc Natl Acad Sci USA 1993; 90:6547-6551.
97. Edgworth J, Freemont P, Hogg N. Ionomycin-regulated phosphorylation of the myeloid calcium-binding protein p14. Nature 1989; 342:189.
98. Nemoto Y, Ikeda J, Katoh K et al. R_2D_5 antigen: a calcium-binding phosphoprotein predominantly expressed in olfactory receptor neurons. J Cell Biol 1993; 123:963-976.
99. Mailleux P, Halleux P, Verslijpe M et al. Neuronal localization in the brain of the messenger RNA encoding calcyphosine, a new calcium-binding protein. Neurosci Lett 1993; 153:125-130.
100. Sauter A, Staudenmann W, Hughes GJ et al. A novel EF-hand Ca^{2+}-binding protein from abdominal muscle of crustaceans with similarity to calciphosine from dog thyroidea. Eur J Biochem 1995; 227:97-101.
101. Klee CB. Ca^{2+}-dependent phospholipid- (and membrane-) binding proteins. Biochemistry 1988; 27:6645-6652.
102. Crompton MR, Moss SE, Crumpton MJ. Diversity in the lipocortin/calpactin family. Cell 1988; 55:1-3.
103. Swairo MA and Seaton BA. Annexin stucture and membrane interactions: a molecular perspective. Annu Rev Biophys Biomol Stucture 1994; 23:193-213.
104. Moss SE, Edwards HC, Crumpton MJ. Diversity in the annexin family. In: Heizmann CW, ed. Novel Calcium Binding-Proteins. Fundamentals and Clinical Implications. Berlin, Springer-Verlag, 1992:535-566.
105. Sato EF, Tanaka Y, Edashige K et al. Annexin I and its biochemical properties. In: Heizmann CW, ed. Novel Calcium-Binding Proteins. Fundamentals and Clinical Implications. Berlin, Springer-Verlag, 1992:567-587.
106. Cirino G, Flower RJ. The anti-inflammatory activity of human recombinant lipocortin I. In: Heizmann CW, ed. Novel Calcium-Binding Proteins. Fundamentals and Clinical Implications. Berlin, Springer-Verlag, 1992:589-611.
107. The Annexins. Moss SE, ed. London and Chapel Hill, Portland Press, 1992.

108. Raynal P, Pollard HB. Annexins: the problem of assessing the biological role for a gene family of multifunctional calcium- and phospholipid-binding proteins. Biochim Biophys Acta 1994; 1197:63-93.
109. Smith PD, Moss SE. Structural evolution of the annexins supergene family. Trends Genetics 1994; 10:241-246.
110. Gerke V. p11, a member of the S-100 protein family, is associated with the tyrosine kinase substrate p36 (annexin II). In: Heizmann CW, ed. Novel Calcium-Binding Proteins. Fundamentals and Clinical Implications. Springer-Verlag, Berlin, 1991:139-155.
111. Jeng F-Y, Gerke V, Gabins H-J. Identification of annexin II, annexin VI and glyceraldehyde-3-phosphate dehydrogenase as calcyclin-binding proteins in bovine heart. Int J Biochem 1993; 25:1019-1027.
112. Watanabe M, Ando Y, Tokumitsu H et al. Binding site of annexin XI on the calcyclin molecule. Biochem Biophys Res Commun 1993; 196:1376-1382.
113. Rothwell NJ, Relton JK. Involvement of interleukin-1 and lipocortin-1 in ischemic brain damage. Cerebrovasc Brain Metab Rev 1993; 5:178-198.
114. Flower RJ, Rothwell NJ. Lipocortin-1: cellular mechanisms and clinical relevance. Trends Pharmacol Sci 1994; 15:71-77.

CHAPTER 3

GENE STRUCTURES AND CHROMOSOMAL ASSIGNMENTS OF EF-HAND CA²⁺-BINDING PROTEINS

The genes of only a few members of the EF-hand family have been analyzed.[1,2] Based on these data it is suggested that these proteins have evolved from a single ancestral EF-hand motif. During evolution multiple reiterations of this primordial gene gave rise to genes coding for proteins with multiple EF-hands. Subsequent gene duplication led to the large family of proteins with different expression patterns and functions.

Generally the genes coding for the various Ca²⁺-binding proteins, even of the same subfamily, are located on different chromosomes (Table 3.1). Recently six S100 genes were detected that are clustered on chromosome 1q21.[3] By isolating a yeast artificial chromosome (YAC) containing the 1q21 region as many as nine human S100 genes were detected (Fig. 3.1 and ref. 4) and a new nomenclature was proposed for the S100 genes located on chromosome 1q21[4] (Table 2.1). The names of S100 genes localized on other chromosomes were not changed (Table 2.1).

The nine human S100 genes on the YAC clone were in the following order: S100A1 (S100A) - S100A2 (S100L) - S100A3 (S100E) - S100A4 (CAPL) - S100A5 (S100D) - S100A6 (CACY) - S100A7 (PSOR1) - S100A8 (CAGA) and S100A9 (CAGB).

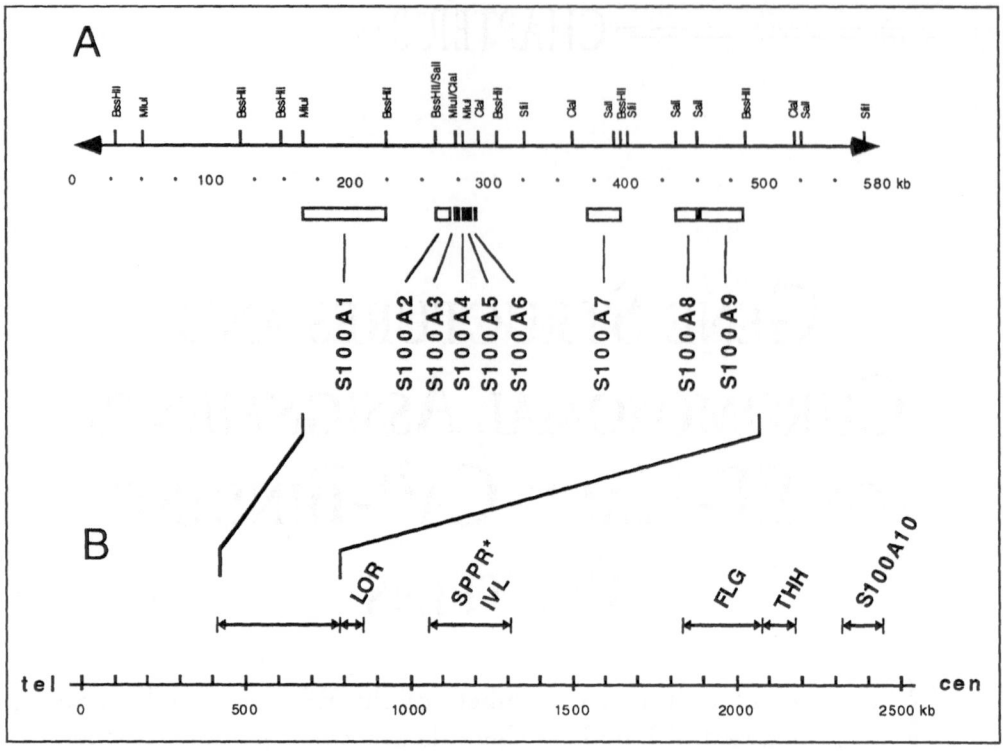

Fig. 3.1. Physical map of the gene cluster on chromosome 1q21. The map of the YAC clone 100F3[4] (A) is linked to the recently published[5] linkage map of the epidermal differentiation genes loricrin (LOR), small proline-rich protein (SPPR), involucrin (IVL), profilaggrin (FLG) and trichohyalin (THH). (B) asterix indicates that SPRR represents a gene family of at least 10 genes located within 300 kb of each other.[14] The relative order of the S100 gene cluster within chromosome 1 (tel/cen) is taken from Hardas et al.[15] In A, S100A3 - S100A6 represent genes that are completely characterized and sequenced (designated by filled boxes). Open boxes indicate genes that are either not completely sequenced (S100A1, S100A2, S100A7) and/or whose exact location is not known (S100A8 and S100A9). Reprinted with permission from: Genomics 25 1995; 638-643, © Academic Press.

A cluster of epidermal differentiation genes has been mapped to the same 1q21 region.[5] In this map, S100A6 was localized to within 450 kb of the gene for the cornified envelope protein loricrin. Furthermore, S100A10 was mapped to a position about 2000 kb distant from S100A6. These two analyses extended the map and it was possible to localize at least 10 S100 genes clustered on human chromosome 1q21.[4]

Interestingly, the genes coding for profilaggrin (FLG) and trichohyalin (THH) appear to be fused genes containing an S100-like domain at the 5' end. This S100 motif seems to be cleaved off upon maturation of the epidermal proteins;[6,7] but it is not currently known if the cleaved S100 proteins are stable. At least 12 S100 genes (including $S100A_1$ to $S100A_{10}$; FLG and THH) must have been duplicated and kept together during evolution. Three additional S100 genes, S100B, calbindin 9K and S100P, are found dispersed in the genome (Table 2.1). During the studies with S100 genes on chromosome 1q21, two novel S100 genes (S100A3 and S100A5) were identified.[3] Since the gene density in this region is very high, the existence of additional novel S100 genes in this cluster cannot yet be excluded.

An interesting question concerns the regulation of the individual genes within the cluster. Most simply, each gene could be controlled by its own individual elements. On the other hand, some of the clustered genes may share superior locus regulatory control elements as has been shown for the globin locus. Indeed, such a possibility has been suggested for the epidermal differentiation genes, since their expression seems to be spatially and temporally coordinated in the human skin.[5] By contrast, no coordinate expression is seen for the S100 genes, where each gene is expressed in a specific subset of tissues or cells (reviewed in Hilt and Kligman[8]). It remains to be seen whether these S100 genes are therefore regulated by separate elements.

The discovery of the S100 gene locus provided a logical basis to simplify the presently used but confusing names of this protein family. The names can now also be used to designate the corresponding S100 genes in different species, using the common one or two letter abbreviations to indicate the species. Since it is known that S100A4, S100A6, and S100A10 are also linked in a syntenic region on mouse chromosomes[9-11] it is possible that the clustered organization of S100 genes is conserved during evolution. At present it is not known whether S100 genes exist in the genome of invertebrates.

The finding of physical linkage of genes that are evolutionarily closely related appears to be more than coincidental. In fact, there are several examples of other genes that are highly conserved during evolution and interact functionally (e.g. globin gene cluster, major histocompatibility complex, and Hox genes) and also retain a close physical linkage. It will be of great interest to see if S100 genes are clustered in other species, as well.

Table 3.1. *Chromosomal assignment of the genes encoding human EF-hand calcium-binding proteins*

Protein	Human chromosome	Selected references
α-Parvalbumin	22q12-q13.1	1,2
β-Parvalbumin (oncomodulin)	7	2
Calbindin D-28K	8q21.3-q22.1	3
Calretinin	16q22-q23	3
Recoverin	17p12-p13	4,5
Hippocalcin	1	6
Hippocalcin-like protein (hHLP$_2$)	2	7
Calmodulin related genes		
CALM 1	14 q24-q31	8,18
CALM 2	2p21.1-p21.3	8,18
CALM 3	19q13.2-q13.3	8,18
Pseudogene hCE$_2$	17	9
Calmodulin-like 1	7pter-p13	10
CALM L	10p13-ter	11
Calpain related genes		
CAN PL1	11	12
CAN PL2	1	12
CAN PL3	15	12
CAN PS	19	12
Calcineurin (neural, catalytic subunit)	4 and 10	13
Calcineurin (testis, catalytic subunit)	8	13
Sorcin	7	14
Sorcin-related gene	4	14
Myosin essential light-chain		
MLC-1$_F$, MLC-3$_F$	2q32.1-qter	15
MLC-1emb./A	17q21-qter	16
MYL$_3$/MYL$_4$	3	17
S100 proteins	see Table 2.1	

1. Berchtold MW, Epstein P, Beaudet AL et al. J Biol Chem 1987; 262:8696-8701.
2. Ritzler JM, Sahwney R, Geurts van Kessel AHM et al. Genomics 1992; 12:567-572.
3. Parmentier M, Passage E, Vassart G et al. Cytogenet Cell Genet 1991; 57:41-43.
4. Wiechman AF, Akots G, Hammarback JA et al. Invest Ophthalm & Vis Sci 1994; 35:325-331.
5. Murakami A, Yajima T, Inana G. Biochem Biophys Res Commun 1992; 187:234-244.
6. Takamatsu K, Kobayashi M, Saitoh S et al. Biochem Biophys Res Commun 1994; 200:606-611.
7. Kobayashi M, Takamatsu K, Fujishiro M et al. Biochim Biophys Acta 1994; 1222:515-518.
8. McPherson JD, Hickie RA, Wasmuth JJ et al. Cytogenet Cell Genet 1991; 58:1951
9. SenGupta B, Detera-Wadleigh SD, McBride OW et al. Nucleic Acid Res 1989; 17:2868.
10. Scambler PJ, McPherson MA, Bates G et al. Human Genet 1987; 76:278-282.
11. Berchtold MW, Koller M, Egli R et al. Human Genet 1993; 90:496-500.
12. Ohno S Minoshima S, Kudoh J et al. Cytogenet Cell Genet 1990; 53:225-229.
13. Maramatsu T, Kincaid RL. Biochem Biophys Res Commun 1992; 188:265-271.
14. Van der Bliek Am, Baas F, Van der Velde-Koerts T et al. Cancer Res 1988; 48:5927-5932.
15. Cohen-Hagenauer O, Barton PJR, van Cong N et al. Human Genet 1988; 78:65-70.
16. Seharaseyon J, Bober E, Hsieh C-L et al. Genomics 1990; 7:289-293.
17. Cohen-Hagenauer O, Barton PJR, van Cong N et al. Human Genet 1989; 81:278-282.
18. Berchtold MW, Egli R, Rhyner JA et al. Genomics 1993; 16:461-465.

Interestingly, alterations involving the chromosomal region of 1q21 are seen in a variety of preneoplastic and neoplastic diseases[12,13] (chapter 6) and occur as an early event in some cases. Therefore it will be interesting to examine to what extent S100 genes are involved in tumongenesis.

REFERENCES

1. Heizmann CW, Hunziker W. Intracellular calcium-binding proteins: more sites than insights. Trends Biochem Sci 1991; 16:98-103.
2. Berchtold MW. Evolution of EF-hand calcium-modulated proteins. V. The genes encoding EF-hand proteins are not clustered in mammalian genomes. J Mol Evol 1993; 36:489-496.
3. Engelkamp D, Schäfer BW, Mattei MG et al. Six S100 genes are clustered on human chromosome 1q21: Identification of two genes coding for the two previously unreported calcium-binding proteins S100D and S100E. Proc Natl Acad Sci USA 1993; 90:6547-6551.
4. Schäfer BW, Wicki R, Engelkamp D et al. Isolation of a YAC clone covering a cluster of nine S100 genes on human chromosome 1q21: rationale for a new nomenclature of the S100 calcium-binding protein family. Genomics 1995; 25:638-643.
5. Volz A, Korge BP, Compton JG et al. Physical mapping of a functional cluster of epidermal differentiation genes on chromosome 1q21. Genomics 1993; 18:92-99.
6. McKinley-Grant LJ, Idler WW, Bernstein IA et al. Characterization of a cDNA clone encoding human filaggrin and localization of the gene to chromosome region 1q21. Proc Natl Acad Sci USA 1989; 86:4848-4852.
7. Lee SC, Kim IG, Marekov LN et al. The structure of human trichohyalin: Potential multiple functions as an EF-hand-like calcium binding protein, a cornified cell envelope precursor and an intermediate filament associated (crosslinking) protein. J Biol Chem 1993; 268:12164-12176.
8. Hilt DC, Kligman D. The S-100 protein family: a biochemical and functional overview. In: Heizmann CW, ed. Novel Calcium-Binding Proteins. Berlin: Springer Verlag, 1991:65-103.
9. Saris C, Kristensen T, D'Eustachio P et al. cDNA sequence and tissue distribution of the mRNA for bovine and murine p11, the S100-related light chain of the protein-tyrosine kinase substrate p36 (calpactin I). J Biol Chem 1987; 262:10663-10671.
10. Moseley WS, Seldin MF. Definition of mouse chromosome 1 and 3 gene linkage groups that are conserved on human chromosome 1: Evidence that a conserved linkage group spans the centromere of human chromosome 1. Genomics 1989; 5:899-905.
11. Dorin JR, Emslie E, Van Heyningen V. Related calcium-binding proteins map to the same subregion of chromosome 1q and to an

extended region of synteny on mouse chromosome 3. Genomics 1990; 8:420-426.

12. Craig RW, Jabs EW, Zhou P. et al. Human and mouse chromosomal mapping of the myeloid cell leukemia-1 gene: MCL1 maps to human chromosome 1q21, a region that is frequently altered in preneoplastic and neoplastic disease. Genomics 1994; 23:457-463.

13. Pedrocchi M, Schäfer BW, Mueller H. et al. Expression of Ca^{2+}-binding proteins of the S100 family in malignant human breast-cancer cell lines and biopsy samples. Int J Cancer 1994; 57:684-690.

14. Gibbs S, Fijmenam R, Wiegant J et al. Molecular characterization and evolution of the SPRR family of keratinocyte differentiation markers encoding small proline-rich proteins. Genomics 1993; 16:630-637.

15. Hardas BD, Zhang J, Trent JM et al. Direct evidence for homologous sequences on the paracentric regions of human chromosome 1. Genomics 1994; 21:359-363.

CHAPTER 4

LOCALIZATION OF EF-HAND Ca²⁺-BINDING PROTEINS IN THE CNS

T he intention of this chapter is not to provide a complete review of all the anatomical literature dealing with the distribution of Ca²⁺-binding proteins in the central nervous system, rather an attempt is made to survey the distribution and cellular localization of Ca²⁺-binding proteins in selected brain systems of the mammalian brain. Special emphasis is placed on brain systems that are affected in certain neurodegenerative diseases of the human brain, such as hippocampus, cerebellum, basal ganglia and cortical areas. Since, for most of these brain regions, excellent, extensive, and very detailed descriptions of cell types, their distribution and pathways containing different Ca²⁺-binding proteins have already been published (for reviews see refs. 1-6), the following description is restricted to brain areas and their neurons and functional pathways, which are related to the pathology of brain diseases.

HIPPOCAMPUS

The mammalian hippocampus contains neuronal populations characterized by their immunoreactivity with antisera against parvalbumin, calbindin D-28K, calretinin, calcineurin and calmodulin. In both, rodents and primates, similar populations of principal neurons contain calbindin D-28K, and the two proteins are localized in distinct and non-overlapping populations of non-principal neurons which show similar target selectivity in all species investigated. Antibodies against calbindin D-28K stain various cells, including a large proportion of the pyramidal cells of fields CA1 and CA2, some interneurons and virtually all granule cells of the dentate gyrus together with their mossy fiber projection to the CA3 region.[1,7-10] In human hippocampus field CA2 contains the highest density of calbindin D-28K-immunoreactive pyramidal neurons and their abrupt disappearance clearly marks its border with the CA3 area.[10] The differences in density and staining intensity of the pyramidal cells in CA1 and CA2 is less pronounced.

The distribution, morphological features and synaptic connections of neurons containing parvalbumin and calbindin D-28K in primate hippocampus has been intensively studied[10-15] and these studies revealed that the human hippocampus displays the largest variability of these immunoreactive neurons both in their morphology and location. Despite some controversial reports concerning the presence of calbindin D-28K in CA1 pyramidal neurons of different primates,[8,9] large numbers of CA1 pyramidal neurons in the rhesus and baboon hippocampus as well as in the human hippocampus contain calbindin D-28K.[9] A distinct population of calbindin D-28K-positive local circuit neurons was found in all layers of the dentate gyrus and the pyramidal layer of Ammon's horn. In human hippocampal formation, calbindin D-28K-immunoreactive non-pyramidal neurons frequently occur in the molecular layer of the dentate gyrus and in the stratum lacunosum-moleculare of Ammon's horn, whereas in rat or monkey hippocampus these regions are devoid of such neurons.[15] In the subicular complex the pyramidal neurons are not calbindin D-28K-immunoreactive. In the prosubiculum and subiculum calbindin D-28K-immunoreactive non-pyramidal neurons are distributed in all layers, in the presubiculum they occur in the superficial layers. In primates the CA2 subfield shows some characteristics with regard to calbindin D-28K- and parvalbumin-immunoreactive interneurons: only very few calbindin D-28K-positive interneurons are found in the monkey CA2 region,[7] whereas a very high density of parvalbumin-positive interneurons and their terminals surround CA2 pyramidal neurons.[7,13] Parvalbumin-immunoreactivity in the human hippocampus is present exclusively in local circuit neurons in the hippocampal formation and they are equally distributed in all layers of the subicular complex.[14] Most of the parvalbumin-positive neurons are located close to the principal cell layers but large numbers of immunoreactive neurons were also present in the molecular layer of the dentate gyrus and in stratum oriens of Ammon's horn. In the human hippocampus CA2 is characterized by the absence of parvalbumin-positive interneurons from strata radiatum and moleculare and thereby is distinguished from CA3 and CA1 regions.[11] Parvalbumin-positive neurons are basket cells and axo-axonic cells and they display radially oriented dendritic trees spanning all layers.[1,16-20]

HIPPOCAMPAL NON-PYRAMIDAL CELLS, INHIBITORY INTERNEURONS

In rodents calbindin D-28K- and parvalbumin-containing local circuit neurons are GABAergic,[16,21] they form non-overlapping subpopulations and differ in target selectivity and in afferent innervation.[14,20] There seems to be a direct relationship between the content of a given Ca²⁺- binding protein and postsynaptic target selection of afferent input. For instance, the neurons innervating the perisomatic region of pyramidal cells (basket and axo-axonic cells) are parvalbumin-immunoreactive,[17,18] whereas those terminating in the dendritic layers and making synapses on dendritic shafts or spines contain calbindin D-28K or calretinin. The GABAergic parvalbumin-immunoreactive axo-axonic cells show the most precise target selectivity of non-pyramidal cells, they form synapses exclusively with axon initial segments of principal cells in the hippocampus and dentate gyrus.[22-26] The parvalbumin-immunoreactive interneurons in the CA3 region receive highly convergent excitatory input via single synapses on their dendrites from several CA3 pyramidal neurons, and it has been shown that one pyramidal neuron projects in a highly divergent manner onto many parvalbumin-immunoreactive neurons.[27] Since the GABAergic/parvalbumin-immunoreactive interneurons are known to express muscarinic receptors they are assumed to receive excitatory input from the medial septum and diagonal band.[28,29] The basket cells, which contain the inhibitory transmitter GABA,[23] can be further subdivided into three cell classes on the basis of their axonal arborizations restricted to the cell body region of pyramidal neurons.[30] Some of these do not respect the boundaries of hippocampal subfields, i.e. their axonal arbors cross the CA3/CA2/CA1 boundaries or may innervate dentate granule cells as well as hilar and CA3 neurons.[31] This suggests that there is no complete separation of inhibitory circuits.[30]

Calretinin-immunoreactivity is present exclusively in non-granule and non-pyramidal neurons in all layers of the dentate gyrus and in the CA1-CA3 areas in the hippocampal formation of rats and monkeys.[14,30,32] In rats calretinin-immunoreactive neurons are most abundant in the hilus of the dentate gyrus and in stratum lucidum of CA3 where they can be subdivided into two distinct cell groups.[33] An aspiny neuron type is found in all areas and layers of the hippocampus, a spiny neuron type is found exclusively in the stratum lucidum of the CA3 area and in the hilar region of the dentate gyrus. The spiny calretinin-immunoreactive neurons in stratum lucidum of CA3 and in the hilus display a remarkable input specificity receiving an almost exclusive and highly convergent input from asymmetrical synapses of the mossy fibers.[33,34] The aspiny calretinin-positive neurons show the characteristic features of GABAergic interneurons, their varicose axons forming symmetric contacts on dendritic shafts and sometimes on somata in all layers of the CA3 region. Colocalization studies revealed that the majority of calretinin-immunoreactive neurons are GABA-positive in all regions of the hippocampal formation and that the two morphologically distinct neuron types differ in their immunoreactivity for GABA.[34] Almost all aspiny calretinin-positive neurons are GABA-positive whereas most of the spiny calretinin-immunoreactive neurons were GABA-negative. The spiny neuron type in the hilus seems not to exist in monkeys.[32] Furthermore, in rat and monkey hippocampus there is no colocalization of calretinin and the two other Ca^{2+}-binding proteins, parvalbumin and calbindin D-28K.[14,32,34]

In both rats and monkeys a distinct band of dense calretinin-immunoreactive neuropil is observed in the superficial region of the granule cell layer and in the lowest part of the molecular layer. The characteristic location and ultrastructural morphological features of these synapses, which are mostly asymmetric, suggests that they derive from extra-hippocampal afferents.[33] In green monkeys it was shown that they represent substance P-immunoreactive fibers arising from the supramammillary nucleus of the hypothalamus.[32] Thus, in monkey hippocampal formation there seem to be at least two separate calretinin-containing systems: an intrinsic inhibitory system and an extrinsic excitatory system,[32] and in the hilus an additional system of a few large spiny neurons project via the commissure to the contralateral hippocampus.[34]

In summary, although there are parvalbumin-, calbindin D-28K- and calretinin-positive cell types which are unique to the human hippocampus, these Ca²⁺-binding proteins are located in similar populations of neurons in human, non-human primates and in rodents. Despite a few exceptions general pattern and distribution of these neuron populations is very similar and this encourages the use of rodent models for studying the pathological mechanisms of epilepsy and ischemia.[14,19,35,36]

Calmodulin-immunoreactive cells are distributed in the stratum granulosum and stratum pyramidale and in the subiculum.[37,38] Calcineurin-positive neurons can be found particularly in the strata pyramidale radiatum and lacunosum, while the dentate gyrus displays moderate calcineurin-immunostaining.[39] In the hippocampus proper and in the dentate gyrus calcineurin is precisely colocalized in almost all neurons with the type II Ca²⁺/calmodulin-dependent protein kinase (calmodulin-kinase II), suggesting a close interaction of the phosphatase calcineurin and the calmodulin-kinase II in such neurons.[40]

Another Ca²⁺-binding protein, chromogranin A, has been demonstrated in CA2 pyramidal neurons of the human hippocampus.[41]

STRIATUM/BASAL GANGLIA

Similar to the hippocampal formation and many other brain regions, parvalbumin and calbindin D-28K are expressed in different neuron populations in the basal ganglia, and the regional localization of these neuron populations in the striatum is in general terms largely complementary. In the human brain parvalbumin-immunoreactivity is mainly present in the more dorsal regions of the dorsal striatum whereas calbindin D-28K-immunoreactivity is concentrated in the ventral striatum and the adjacent ventral and medial regions of the dorsal striatum.[42] Parvalbumin-immunoreactivity is found in some medium-sized aspiny GABAergic interneurons[43-45] most of which are located in the matrix compartment, some at the border of the patches and a few within the patches.[46-48] Parvalbumin-immunoreactive neurons receive direct cortical inputs and innervate the calbindin D-28K-immunoreactive spiny projecting neurons.[44] In the human striatum the compartmentalization of parvalbumin-immunoreactive neurons in the caudate-putamen complex in principle follows the same compartmental mosaic as described for other mammals.[46,47,49] Here, the parvalbumin-poor patches align precisely with the enkephalin-rich striosomes and conversely, the parvalbumin-rich region corresponds to the matrix zone. The calbindin D-28K-immunoreactive neurons in the caudate-putamen complex show a gradient with weakly immunoreactive neurons and neuropil in the dorsolateral striatum and dense, heavily calbindin D-28K-immunoreactive neurons and fibers in the matrix compartments.[48] These medium-sized spiny calbindin D-28K-positive neurons represent a subset of the striato-nigral GABAergic/glutamatergic projection neurons that project to the substantia nigra pars reticulata,[50] which contains parvalbumin-positive neurons. A small percentage of the calbindin D-28K-positive neurons in the patches also contain neuropeptide Y[48] and the aspiny calbindin D-28K-positive neurons express NADPH-diaphorase.[51,52] The dorsal tier of the substantia nigra pars compacta contains many calbindin D-28K-immunoreactive neurons, which are assumed to be dopaminergic and which project to the striatal matrix,[47,53] whereas the dopaminergic neurons in the ventral tier of the substantia nigra pars compacta are calbindin D-28K-negative and project to the striatal patches. The calbindin D-28K-immunoreactive neurons in the dorsal tier are selectively spared from degeneration after 6-hydroxydopamine lesions during development of the mesostriatal dopaminergic system.[47,53]

The small population of calretinin-immunoreactive neurons in the striatum of rats represents another subpopulaton of medium-sized aspiny interneurons, they contain glutamate decarboxylase but they are distinct from neurons containing parvalbumin, choline acetyltransferase, NADPH-diaphorase or nitric oxide synthase.[51,54] Their distribution shows a conspicuous rostro-caudal gradient with a greater density of calretinin-positive neurons in the rostral parts of the striatum.

Calmodulin- and calcineurin-immunoreactivity show identical profiles in the caudate-putamen complex.[55,56] Large somata along with their processes prove to be calmodulin- and calcineurin-positive. Some of the long calmodulin-immunoreactive fibers are observed to enter the corpus callosum.[56] The myelinated fibers of the white matter of the caudate-putamen complex are negative in adult mice quite in contrast to the clear calmodulin-staining manifested in young animals. In the rat, most neurons in the caudate-putamen complex show intense calcineurin-immunoreactivity in soma and neurites, the dorso-lateral part of this region showing the strongest staining. The fiber bundles within the white matter are slightly stained. Biochemically it has been shown that the striatum belongs to the brain regions with the highest calmodulin content together with the hippocampus and the amygdala[57] and a decrease of calmodulin content has been biochemically evaluated in the striatum of rats during aging.[58,59]

CORTEX

Parvalbumin-, calbindin D-28K- and calretinin-immunoreactive neurons belong to principally non-overlapping subpopulations of GABAergic interneurons in the rat, cat, monkey and human cortex[1,60-72] (Compare also Fig. 4.1A-C: presumptive cortical analogue in chick brain). Parvalbumin-positive neurons are abundant in all layers except layer I, they resemble non-spiny stellate cells, chandelier cells, double-bouquet cells and basket cells.[66,68-71] They essentially comprise all non-pyramidal cells and mainly multipolar cells, with the exception of visual cortical areas, where parvalbumin is transiently expressed in pyramidal cells during early postnatal development[72,73] and auditory cortex, where lightly parvalbumin-immunoreactive pyramidal projection neurons are observed.[74] Calbindin D-28K-immunoreactive neurons form another subset of non-pyramidal neurons in the neocortex, and they include the double-bouquet cells, small multipolar neurons and bitufted cells.[63,69,75,76] The intensively labeled neurons are largely confined to layer II, and additionally, pyramidal neurons in layers III and V are lightly calbindin D-28K-positive.[72,74,77,78] This distribution pattern was confirmed in a study using the in situ hybridization technique in mouse occipital, parietal and frontal cortices, which revealed a dense band of strongly labeled calbindin D-28K mRNA-containing neurons in layers II-III and many intensely labeled scattered neurons throughout layers IV and VI.[79] In brains of Alzheimer patients, the heavily labeled neurons seem to be resistant to degeneration, whereas the lightly stained neurons seem to undergo degenerative alterations (see chapter 6).

In the neocortex and in area 17 of *Macaca fasciculata*, parvalbumin- and calbindin D-28K-immunoreactivity is found in separate neuron populations of GABA-positive neurons and affer-

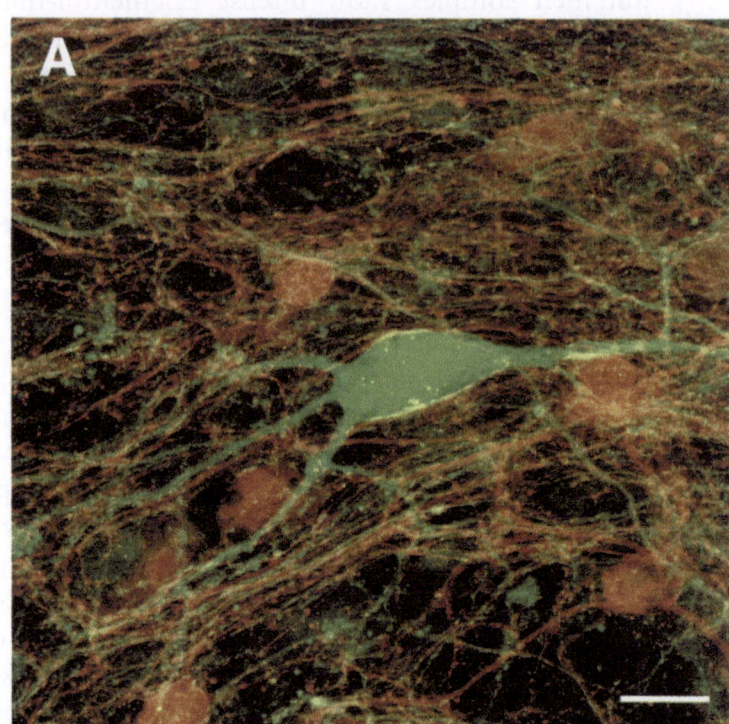

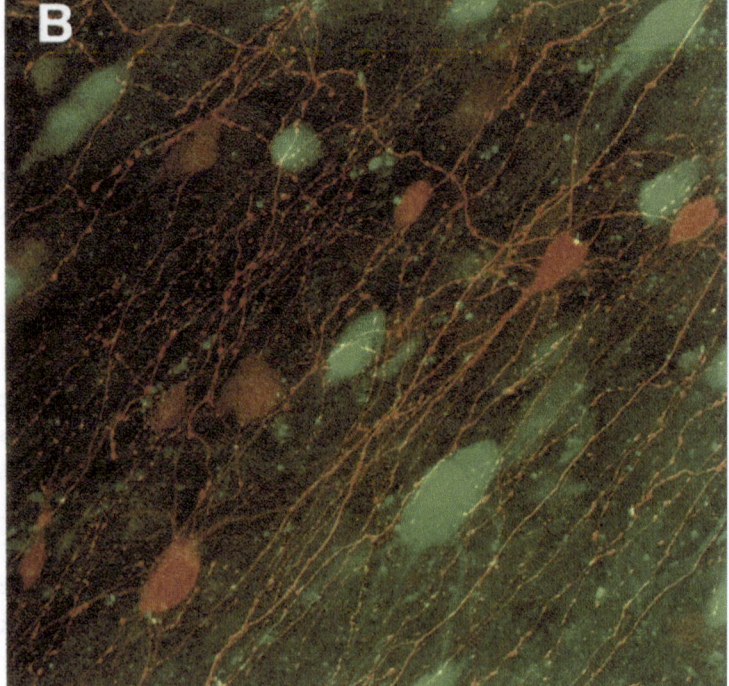

Fig 4.1. Double immunofluorescence labeling for parvalbumin, calbindin D-28K and calretinin in organotypic slice cultures of a learning-relevant associative forebrain region in the domestic chick.

Confocal microscopy of neurons and neuropil demonstrates that the three Ca²⁺-binding proteins label separate neuron populations.

A) The large green calbindin D-28K-immunoreactive soma is surrounded by dense green calbindin D-28K-immunoreactive and red parvalbumin-immunoreactive fibers. The red parvalbumin-immunoreactive neurons are small to medium sized. Calibration bar is 20 μm and corresponds also for B.

B) The red small to medium sized calretinin-immunoreactive neurons develop extensive varicose dendrites and axonal arborizations, which appear to contact the large green calbindin D-28K-immunoreactive somata.

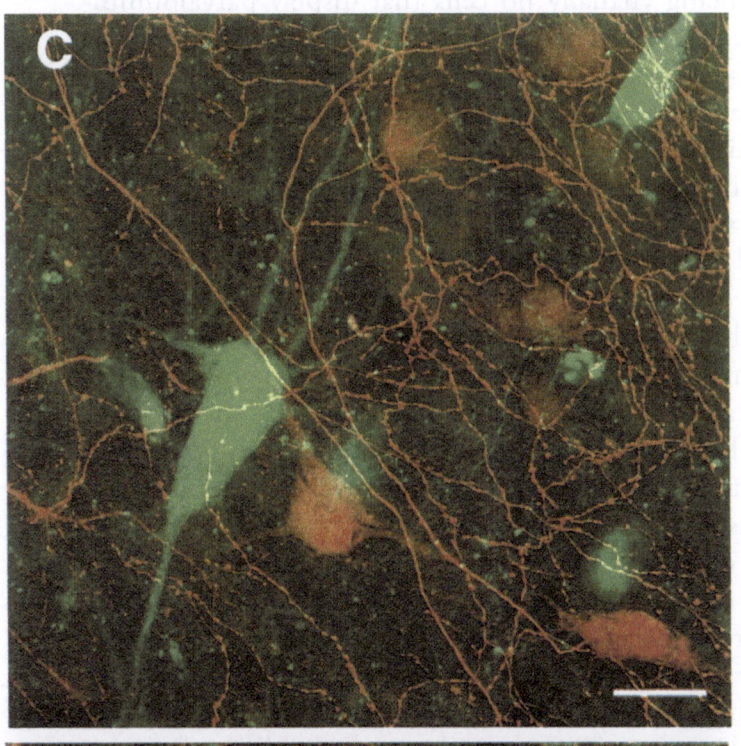

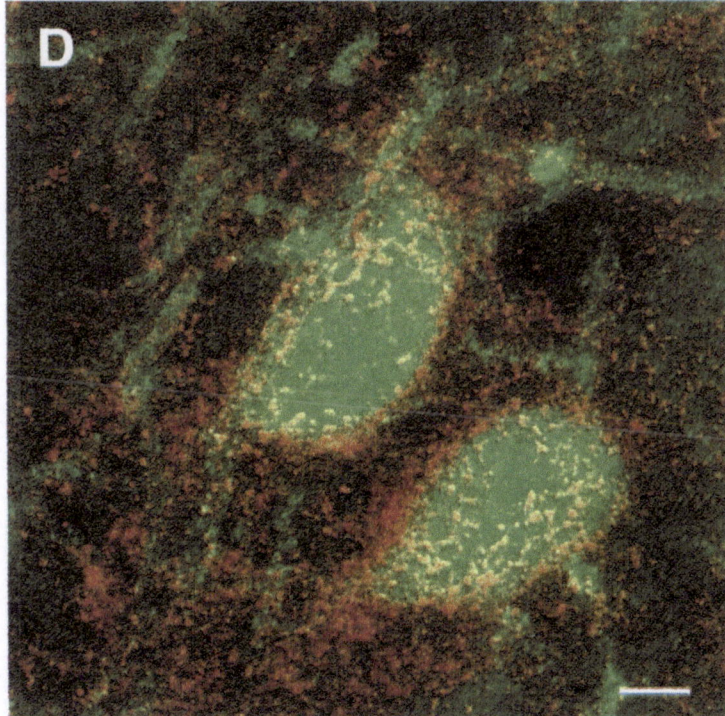

Fig. 4.1. (continued)
C) The dense varicose processes of small to medium sized red calretinin-immunoreactive neurons surround the large green parvalbumin-immunoreactive neurons. Bar is 20 μm
D) Large green parvalbumin-immunoreactive somata are surrounded by red tenascin-immunoreactivity. Bar is 5 μm. (Courtesy of Dr. Martin Metzger, Federal Institute for Neurobiology, Madgeburg)

ent axons, there being virtually no cells that display parvalbumin-plus calbindin D-28K-immunoreactivity. Most parvalbumin- and calbindin D-28K-positive cells also display GABA-immunoreactivity and only a relatively small number of GABA-immunoreactive cells lack immunoreactivity for both Ca^{2+}-binding proteins. Parvalbumin-positive neurons predominate in layers IIIB and IVA and IVC, whereas calbindin D-28K-immunoreactive neurons are primarily located in layers II and IIIA.[63,64,67,68,72,74-78]

A comparative study of neurons containing Ca^{2+}-binding proteins in mammalian cortex revealed species-specific differences of their distribution in different cortical layers.[80-82] A comparison of the compartmental organization of cytochrome oxidase-rich regions with the distribution of calbindin D-28K-immunoreactive GABAergic interneurons in area 17 of squirrel monkeys, macaques and adult humans revealed some species-specific differences in staining patterns and compartmentalization schemes, which may reflect subtle variations of certain neuronal properties between these species.[83] The general pattern for layer III is similar in all primate species, here the calbindin D-28K-positive neurons form a well stained matrix that surrounds the cytochrome oxidase-rich puffs. However, in the human striate cortex there is an additional prominent band of lightly stained calbindin D-28K-positive neurons in layer IVA, which is also observed in squirrel monkeys but not in macaques.

In primate visual cortex and in human temporal cortex parvalbumin-immunoreactive terminal boutons, presumably representing the non-GABAergic geniculostriate projection, form prominent bands in layers IV and IIIB,[70,72,84,85] which correlates spatially with bands of high cytochrome oxidase activity.[67] In contrast, calbindin D-28K-immunoreactive neurons are located in areas of low cytochrome oxidase activities.[62] This distinct localization of the parvalbumin- and calbindin D-28K-immunoreactive structures in areas of high (parvalbumin) or low (calbindin D-28K) metabolic activities, which was first discovered in the avian visual and auditory systems,[86,87] may indicate different activity levels of these two neuronal subpopulations and the pathways in which they are integrated.

Calretinin-immunoreactivity is present in a morphologically distinct population of local circuit neurons in superficial layers in monkey prefrontal cortex,[88] which project to both pyramidal and non-pyramidal neurons. They include the bipolar cells, which are located in the deeper cortical layers[89] and the density of these interneurons is increased in the brain of schizophrenic patients.[90]

The calcium-dependent, calmodulin-stimulated protein phosphatase calcineurin shows a laminar distribution in the adult primary visual cortex in the cat.[91] Calcineurin-immunoreactivity is found mainly in a subpopulation of pyramidal neurons, but non-pyramidal neurons are also stained. Dense labeling is observed in the upper half of layers II/III and in two lightly labeled stripes in lower layer IV and VI. In most pyramidal neurons and also in other neurons in the rat cortex calcineurin is colocalized with the type II Ca^{2+}/calmodulin-dependent protein kinase (calmodulin-kinase II), suggesting a close interaction of the phosphatase and kinase in such neuronal subsets.[92]

The Ca^{2+}-binding protein
   is most abundant in cortical layer II and moderate expression is found in layers II-VI in the rat parietal and occipital cortices.[93] Most of the labeled neurons are pyramidal cells with distinctive staining of the apical dendrites, but a few small to medium sized oval cells are also moderately stained.

S100β protein is expressed in mature and also in immature astrocytes in the visual cortex of cats, with highest levels in layers I-III and V and lowest levels in layer IV.[94] It is localized in the astroglial cytoplasm as well as in the extracellular interstitium. It is also found in oligodendrocytes in the subcortical white matter.

CEREBELLUM
   In contrast to most other brain regions, where parvalbumin, calbindin D-28K and calretinin are mainly represented in separate neuron populations, the principal neuron types of the cerebellum contain more than one of these proteins. Cerebellar Purkinje cells are immunoreactive for calbindin-D-28K,[49,95-99] parvalbumin,[100-102] calcineurin, calmodulin,[103-105] hippocalcin[93] and, in avian brain, also for S100 protein[106] (and our own observations). The presence of either calbindin D-28K- or parvalbumin-mRNA in Purkinje cells has been demonstrated by in situ hybridization, whereas calretinin-mRNA is absent in chick Purkinje cells.[79,107,108] The cerebellar basket and stellate cells contain parvalbumin.[100] Calbindin D-28K-

immunoreactivity was also described in Golgi type II cells[49] and in climbing fibers,[97-99,109] however, this was not confirmed in a study using the in situ hybridization technique, where no calbindin D-28K mRNA was detected in Golgi type II cells.[79,110]

In human cerebellum, calbindin D-28K-immunoreactivity is abundant in Purkinje cells including their axons, dendrites and spines, and it is present in a subpopulation of Golgi type II neurons.[111-113]

A comparative assessment of parvalbumin-content in different brain regions revealed the highest intracellular parvalbumin-concentrations in the cerebellum followed by cortical areas, the hippocampus and the caudate-putamen complex.[114,115]

In rat and chick Purkinje cells, an ATP-dependent plasma membrane Ca^{2+}-pump is colocalized with parvalbumin and calbindin D-28K-immunoreactivity.[116] Particularly within the more distal segments of the Purkinje cell dendrites including spines, a distinctly different distribution pattern of calbindin D-28K and the Ca^{2+}-pump is obvious, with calbindin D-28K/parvalbumin located within the cytoplasm and the Ca^{2+}-pump associated with the cell membrane. There is some evidence that calbindin D-28K directly stimulates the activity of the Ca^{2+}-pump in intestinal and erythrocyte membranes and modulates the activity of voltage-sensitive Ca^{2+}-channels and it remains to be shown whether it serves similar functions in neurons. In Purkinje cells calbindin D-28K-immunoreaction product is closely colocalized with 1,4,5-triphosphate receptors and ryanodine receptors, which are assumed to be associated with intracellular calcium stores.[117]

Calretinin-immunoreactivity has been described in ascending mossy and climbing fibers, in stellate cells but not in Purkinje cells of chicken cerebellum.[118,119] In rat brain intensely calretinin-immunoreactive granule cells and few immunopositive glomeruli were observed, as well as some strongly labeled Lugaro- and Golgi-type II cells.[119] In the molecular layer calretinin-immunoreactive varicose parallel fibers have been described making synaptic contacts with unlabeled Purkinje cell spines.[120] In human and rat cerebellum in addition to the calretinin-immunoreactive fibers, glomerular formations and Golgi-type II and Lugaro cells, a new monodendritic calretinin-immunoreactive neuron type has been discovered in the granule cell layer, which does not contain the

Ca²⁺-binding proteins parvalbumin, calbindin D-28K or chromo-granin C or A.[121-123]

Granule cells also contain the Ca²⁺-binding proteins cal-cineurin[105,106,124-126] and the visinin-like protein VILIP,[127] a few of them also contain calmodulin.[104] In these neurons the calcium-calmodulin-dependent protein phosphatase calcineurin is colocalized with cyclophilin, a family of soluble receptor proteins which bind to the widely used immunosuppressant drug cyclosporin A, and with FK506 binding protein, which binds a pharmacologically simi-lar immunosuppressant.[126] Although there are notable exceptions, the striking similarities in localization of the two immunophilins FK506 binding protein and cyclophilin with calcineurin, which is observed in most brain regions including the hippocampal forma-tion, agree with the observation that immunophilins exert their major biological actions through inhibition of calcineurin activity. Furthermore, these observations indicate some key regulatory func-tion of these mechanisms in the brain. For instance, FK506 and cyclosporin A protect against glutamate-induced neurotoxicity in cortical cultures by preventing the calcineurin-mediated dephos-phorylation of nitric oxide synthase.[128] In addition, FK506 enhances neurotransmitter release by preventing calcineurin-mediated dephos-phorylation of synaptic vesicle proteins.[129]

In the deep cerebellar nuclei, calbindin D-28K- and parval-bumin-immunoreactive punctate structures occur, probably resem-bling the terminal boutons of the Purkinje cell axons,[49,101] as well as calmodulin, calcineurin and S100-immunoreactive somata.[37,104,106]

ULTRASTRUCTURAL LOCALIZATION

The ultrastructural localization and intracellular distribution has been investigated for parvalbumin, calbindin D-28K, calretinin, calmodulin and calcineurin.

For parvalbumin-immunocytochemistry pre- and postembedding staining methods yielded slightly dissimilar results. In the vocal motor nucleus HVc of zebra finches, using the preembedding in-direct peroxidase method, parvalbumin reaction product was lo-cated in amorphous material of perikarya, in some nuclei, in den-drites including spines and in some axons.[130] An association with microtubuli, postsynaptic densities (PSD) and such intracellular membranes as the outer mitochondrial membrane and the endoplasmic reticulum (ER) was observed, whereas the cisternae of the Golgi apparatus and the ER remained free of reaction product. Parvalbumin reaction product was found mainly in Gray type-2 boutons but rarely in type-1 boutons. This distribution pattern had been confirmed in the visual cortex of cats.[65] Here, the immunopositive terminals on parvalbumin-positive and par-valbumin-negative somata were always symmetrical.

Using the immunoferritin labeling of ultrathin frozen sections (non-embedding staining) in rat and mouse cerebellum, par-valbumin-labeling was found dispersed in the cytoplasm and nu-clei of Purkinje cells. The label in the cytoplasm sometimes ap-peared to be clustered, suggesting an association with cytoskeletal structures. Staining of the postsynaptic densities was not observed by this method, but was seen with the PAP-pre-embedding method.[131]

Electron microscopic observations of parvalbumin-immunore-active neurons in the CA1 region of rat hippocampus reveal gap junctions, but these are only found between parvalbumin-positive dendrites and somata.[17,18]

The ultrastructural distribution of calbindin D-28K-reaction product using the pre-embedding PAP-method suggested an asso-ciation with nearly all membranes of Purkinje cell perikarya. The PSDs of dendritic spines showed particularly strong labeling; and in axons there was labeling of microtubules.[132] Similar observations have been made in calbindin D-28K-immunoreactive neurons of the visual cortex of cats.[65] Here, light microscopically detected calbindin D-28K-immunoreactive punctate structures appeared at the EM level with otherwise unlabeled somata as spherical 0.5-1µm large bodies covered by a membrane and occasionally con-taining small vesicles. Using the peroxidase and immunogold la-beling technique in trigeminothalamic interneurons, calbindin D-28K-immunoreaction product was most concentrated in the cell bodies and dendrites, whereas axon terminals were only rarely stained.[133] The predominant site of calbindin D-28K-immunolabel was the matrix of the cytoplasm, and immunolabel was heavily associated with euchromatin within nuclei. Strong labeling of den-drites and spines and also of axon terminals was observed in the spiny interneurons of the basal ganglia.[50] Again the predominant site of calbindin D-28K-label was in the matrix of the cytoplasm and label was associated with the karyoplasm. In the axons immunolabel was associated with neurofilaments, whereas membranes were either sparsely labeled (endoplasmic reticulum, mitochondria) or devoid of label (nuclear envelope and plasmalemma).

The ultrastructural localization and distribution of calretinin-immunoreactivity has been investigated in the rat cerebellum by the immunoperoxidase method.[120] Calretinin-immunoreaction product was found presynaptically in parallel fibers and their varicosities, which make synaptic contacts with dendritic spines of unlabeled Purkinje cells, and it is present in some mossy fiber terminals. Within the immunostained granule cells the reaction product was found in the cytoplasmic matrix of both the perikaryon and primary dendrites and in euchromatin patches of the nucleus.

The distribution of calmodulin-immunoreaction product was investigated in rat Purkinje and granule cells using the post-embedding Fab-peroxidase technique.[103] Calmodulin reaction product was particularly intensive in the nucleus, while in dendrites the cytoplasm was more lightly stained, being localized on free ribosomes and on the rough ER (RER).[104] Purkinje cell axons were lightly calmodulin-immunoreactive. The reaction product was associated with membranes of the Golgi apparatus, the smooth ER (SER), the plasma membrane of cell bodies and their processes. The cisternae of the RER and of the Golgi apparatus remained free of the reaction product. In dendrites calmodulin immunoreactivity was concentrated at the SER on small vesicles and mitochondria. A concentration of calmodulin-immunoreactivity was found at the inner surface of the postsynaptic membrane, in postsynaptic densities and associated with dendritic and axonal microtubules. During development of the cerebellum and the hippocampus, calmodulin undergoes intracellular redistribution.[104] Five to ten days after birth, calmodulin-immunoreactivity is found associated with the organelles of the apical cones of the outgrowing Purkinje cell dendritic trees with little or no staining of the elongated dendrites or of the cell nucleus being evidenced. Later in development calmodulin-immunoreactivity is localized in primary and secondary dendrites, and by 20 days after birth calmodulin-immunoreactivity appears in the nucleus and in the PSD of the spine and shaft synapses.

In the caudate-putamen complex of mouse brain, localization of calmodulin and calcineurin were almost identical within cell somata and dendrites. The staining within the soma was distributed throughout the cytosol and was deposited in cellular organelles, especially on their membranes facing the cytoplasm. The reaction product was consistently greater at the cell periphery than in the perinuclear region. Non-myelinated axons contained calmodulin as well as calcineurin reaction product. Aside from heavy labeling of the PSDs with both proteins, microtubules and mitochondria were stained.[55]

Using the Fab-conjugated peroxidase method, the ultrastructural distribution of calcineurin-immunoreactivity in the striatanigral pathway was investigated. Somata, dendrites and axons icluding their terminal boutons showed very strong immunoreactivity. In some neurons the nucleus appeared to contain the reaction product. Calcineurin-immunoreactivity was distributed throughout the cytoplasm on the outer surface of mitochondria and on microtubules, and was particularly enriched in PSDs.[39] S100β protein, which is restricted to astrocytes, is distributed in the cytoplasm of the soma and processes.[134]

The comparison of these results establishes striking similarities of the intracellular localizations of parvalbumin, calbindin D-28K, calmodulin and calcineurin. In all brain areas and their various different cell types where the ultrastructural distribution of these Ca^{2+}-binding proteins has been investigated, they have been found to be localized throughout the cytosol of the soma, dendrites (including spines) and axons. Furthermore, all Ca^{2+}-binding proteins seem to be associated with microtubules and probably other cytoskeletal elements and with intracellular membranes, such as the outer mitochondrial membrane and the ER; in the case of calmodulin and calcineurin, however, also on the intracellular side of the cell membrane. All proteins are concentrated in the PSDs. In all cases, the cisternae of the Golgi apparatus and of the ER remain free of reaction product.

On the one hand, the similar intracellular distribution of these Ca^{2+}-binding proteins should be considered with some caution, since methodological inadequacies in electron microscopic immunocytochemical techniques are manifold and might be at least in part responsible for the striking similarity of the EM data.

On the other hand, the close association of Ca^{2+}-binding proteins with mitochondria and ER, both of which are Ca^{2+} stores, points to a possible role of these Ca^{2+}-binding proteins in the mechanisms of intracellular Ca^{2+}-regulation. The close association with microtubules and other cytoskeletal elements suggests an involvement in the dis- and rearrangement of the cytoskeleton, which particularly occurs in dendrites and spines. The ultrastructural localization of calmodulin during early postnatal Purkinje cell development suggests that calmodulin may participate in the regulation of microtubule formation during dendritic and axonal development. Its presence in dendrites and in spines may indicate that it also plays a role in the formation of synapses between parallel fibers and dendritic spines. Furthermore, the presence of Ca^{2+}-binding proteins within the PSD implies an involvement of Ca^{2+}-binding proteins in postsynaptic Ca^{2+}-modulated events, which are discussed in connection with synaptic plasticity in the context of learning and memory.[136-138] The presynaptic localization of parvalbumin, calbindin D-28K and especially of calretinin in terminal boutons may indicate additional functions of these proteins in presynaptic mechanisms such as release or re-uptake of transmitters and neuropeptides or axonal growth and sprouting.

REFERENCES

1. Celio MR. Calbindin D28k and parvalbumin in the rat nervous system. Neurosci 1990; 35:375-475.
2. Braun K. Calcium-binding proteins in avian and mammalian central nervous system: localization, development and possible functions. Progr Histochem Cytochem 1990; 21/1:1-64.
3. Heizmann CW, Braun K (1990) Calcium-binding proteins. Molecular and functional aspects. In: Anghileri LJ, ed. The Role of Calcium in Biological Systems. Boca Raton, FL: CRC Press Inc, 1990; 21-65.
4. Heizmann CW, Braun K. Changes in calcium-binding proteins in human neurodegenerative disorders. Trends Neurosci 1992; Vol 17/7:259-264.
5. Baimbridge KG, Celio MR, Rogers JH. Calcium binding proteins in the nervous system. Trends Neurosci 1992; 15:303-308.
6. Andressen C, Blümcke I, Celio MR. Calcium-binding proteins: selective markers of nerve cells. Cell Tissue Res 1993; 271:181-208.
7. Leranth C, Ribak CE. Calcium binding proteins are concentrated in the CA2 field of the monkey hippocampus: A possible key to this region's resistance to epileptic damage. Exp Brain Res 1991; 85:129-136.
8. Sloviter RS, Sollas AL, Barbaro NM et al. Calcium-binding pro-

tein (calbindin-D28k) and parvalbumin immunocytochemistry in the normal and epileptic human hippocampus. J Comp Neurol 1991; 308:381-396.

9. Seress L, Gulyas AI, Freund TF. Pyramidal neurons are immunoreactive for calbindin D28k in the CA1 subfield of the human hippocampus. Neurosci Lett 1992; 138:257-260.

10. Woodhams PL, Celio MR, Ulfig N et al. Morphological and functional correlates of borders in the entorhinal cortex and hippocampus. Hippocampus 1993; 3:303-312.

11. Braak E, Strotkamp B, Braak H. Parvalbumin-immunoreactive structures in the hippocampus of the human adult. Cell Tissue Res 1991; 264:33-48.

12. Ohshima T, Endo T, Onaya T. Distribution of parvalbumin immunoreactivity in the human brain. J Neurol 1991; 238:320-322.

13. Seress L, Gulyas AI, Freund TF. Parvalbumin- and calbindin D28k-immunoreactive neurons in the hippocampal formation of the Macaque monkey. J Comp Neurol 1991; 313:162-177.

14. Seress L, Gulyas AI, Ferrer I et al. Distribution, morphological features, and synaptic connections of parvalbumin- and calbindin D28k-immunoreactive neurons in the human hippocampal formation. J Comp Neurol 1993; 337:208-230.

15. Pitkänen A, Amaral DG. Distribution of parvalbumin-immunoreactive cells and fibers in the monkey temporal lobe: The hippocampal formation. J Comp Neurol 1993; 331:37-74.

16. Kosaka T, Katsumaru H, Hama K et al. GABAergic neurons containing the calcium-binding protein parvalbumin in the rat hippocampus and dentate gyrus. Brain Res 1987; 419:119-130.

17. Katsumaru H, Kosaka T, Heizmann CW et al. Immunocytochemical study of GABAergic neurons containing the calcium-binding protein parvalbumin in the rat hippocampus. Exp Brain Res 1988; 72:347-362.

18. Katsumaru H, Kosaka T, Heizmann CW et al. Gap-junctions on GABAergic neurons containing the calcium-binding protein parvalbumin in the rat hippocampus (CA1 regions). Exp Brain Res 1988b; 72:363-370.

19. Sloviter RS. Calcium-binding protein (calbindin-D28K) and parvalbumin immunocytochemistry: localization in the rat hippocampus with specific reference to the selective vulnerability of hippocampal neuron to seizure activity. J Comp Neurol 1989; 280:183-196.

20. Gulyas AI, Toth K, Danos P et al. Subpopulations of GABAergic neurons containing parvalbumin, calbindin D29k and cholecystokinin in the rat hippocampus. J Comp Neurol 1991; 312:371-378.

21. Toth K, Freund TF. Calbindin D28k-containing nonpyramidal cells in the rat hippocampus: their immunoreactivity for GABA and projection to the medial septum. Neurosci 1992; 49:793-805.

22. Ribak CE, Seress L. Five types of basket cell in the hippocampal dentate gyrus: a combined Golgi and electron microscopic study. J Neurocytol 1983; 12:577-597.
23. Somogyi P, Kisvarday ZF, Martin KAC et al. Synaptic connections of morphologically identified and physiologically characterized large basket cells in the striate cortex of cat. Neurosci 1983; 10:261-294.
24. Somogyi P, Nunzi MG, Smith AD. A new type of specific inter-neuron in the monkey hippocampus forming synapses exclusively with the axon initial segments of pyramidal cells. Brain Res 1983; 259:137-142.
25. Soriano E, Frotscher M. A GABAergic axo-axonic cell in the fascia dentata controls the main excitatory hippocampal pathway. Brain Res 1989; 503:170-174.
26. Li XG, Somogyi JM, Tepper JM et al. Axonal and dendritic arborization of an intracellularly labeled chandelier cell in the CA1 region of rat hippocampus. Exp Brain Res 1992; 90:519-525.
27. Sik A, Tamamaki N, Freund TF. Complete axon arborization of a single CA3 pyramidal cell in the rat hippocampus, and its relationship with postsynaptic parvalbumin-containing interneurons. Europ J Neurosci 1993; 5:1719-1728.
28. van der Zee EA, de Jong GI, Strosberg AD et al. Parvalbumin-positive neurons in rat dorsal hippocampus contain muscarinic actetylcholine receptors. Brain Res Bull 1991; 27:697-700.
29. Van der Zee EA, Luiten PGM. GABAergic neurons of the rat dorsal hippocampus express muscarinic acetylcholine receptors. Brain Res Bull 1993; 32:601-609.
30. Gulyas AI, Miles R, Hajos N et al. Precision and variability in postsynaptic target selection of inhibitory cells in the hippocampal CA3 region. Eur J Neurosci 1993; 5:1729-1751.
31. Han Z-S, Buhl EG, Lörinczi et al. A high degree of spatial selectivity in the axonal and dendritic domains of physiologically identified local-circuit neurons in the dentate gyrus of the rat hippocampus. Eur J Neurosci 1993; 5:395-410.
32. Nitsch R, Leranth C. Calretinin immunoreactivity in the monkey hippocampal formation - II. Intrinsic GABAergic and hypothalamic non-GABAergic systems: an experimental tracing and co-existence study. Neurosci 1993; 55:797-812.
33. Gulyas AI, Miettinen R, Jacobowitz DM et al. Calretinin is present in non-pyramidal cells of the rat hippocampus - III. A new type of neuron specifically associated with the mossy fiber system. Neurosci 1992; 48:1-27.
34. Miettinen R, Gulyas AI, Baimbridge KG et al. Calretinin is present in non-pyramidal cells of the rat hippocampus - II. Coexistence with other calcium binding proteins and GABA. Neurosci 1992; 48:29-43.

35. Kamphuis W, Huisman E, Wadman WJ et al. Kindling induced changes in parvalbumin immunoreactivity in rat hippocampus and its relation to long-term decrease in GABA-immunoreactivity. Brain Res 1989; 479:23-34.
36. Freund TF, Buszaki G, Leon A et al. Relationship of neuronal vulnerability and calcium binding protein immunoreactivity in ischemia. Exp Brain Res 1990; 83:55-66.
37. Seto-Ohshima A, Kitajima S, Sano M et al. Immunohistochemical localization of calmodulin in mouse brain. Histochem 1983; 79:251-257.
38. Seto-Ohshima A. Review: Calcium-binding proteins in the central nervous system. Acta Histochem Cytochem 1994; 27(2):93-106.
39. Goto S, Matsukado Y, Mihara Y et al. The distribution of calcineurin in rat brain by light and electron microscopy immuno-histochemistry and enzyme immunoassay. Brain Res 1986; 397:161-172.
40. Goto S, Matsukado Y, Mihara Y et al. Calcineurin as a neuronal marker of human brain tumors. Brain Res 1986; 371:237-243.
41. Munoz DG. The distribution of chomogranin A-like immunoreactivity in the human hippocampus coincides with the pattern of resistance to epilepsy-induced neuronal damage. Ann Neurol 1990; 27:266-275.
42. Waldvogel HJ, Faull RLM. Compartmentalization of parvalbumin immunoreactivity in the human striatum. Brain Res 1993; 610:311-316.
43. Cowan RL, Wilson CJ, Emson PC et al. Parvalbumin-containing GABAergic interneurons in the rat neostriatum. J Comp Neurol 1990; 302:197-205.
44. Kita H, Kitai ST. Amygdaloid projections to the frontal cortex and the striatum in the rat. J Comp Neurology 1990; 298:40-49.
45. Kita H, Kosaka T, Heizmann CW. Parvalbumin-immunoreactive neurons in the rat neostriatum: a light and electron microscopic study. Brain Res 1990; 536:1-15.
46. Gerfen CR, Baimbridge KG, Miller JJ. The neostriatal mosaic: complementary distribution of calcium-binding protein and parvalbumin in the basal ganglia of the rat and monkey. Proc Natl Acad Sci USA 1985; 82:8780-8784.
47. Gerfen CR, Baimbridge KG, Thibault J. The neostriatal mosaic. III. Biochemical and developmental dissociation of patch-matrix mesostriatal systems. J Neurosci 1987; 7:3935-3944.
48. Kubota Y, Kawaguchi Y. Spatial distributions of chemically identified intrinsic neurons in relation to patch and matrix compartments of rat neostriatum. J Comp Neurol 1993; 332:499-513.
49. Garcia-Segura LM, Baetens D, Roth J et al. Immunohistochemical mapping of calcium-binding protein immunoreactivity in the rat central nervous system. Brain Res 1984; 296:75-86.

50. DiFiglia M, Christakos S, Aronin N. Ultrastructural localization of immunoreactive calbindin-D28k in the rat and monkey basal ganglia, including subcellular distribution with colloidal gold labeling. J Comp Neurol 1989; 279:653-665.
51. Bennett BD, Bolam JP. Characterization of calretinin-immunoreactive structures in the striatum of the rat. Brain Res 1993; 609:137-148.
52. Bennett BD, Bolam JP. Two populations of calbindin D28k-immunoreactive neurons in the striatum of the rat. Brain Res 1993; 610:305-310.
53. Gerfen CR. The neostriatal mosaic: multiple levels of compartmental organization. Trends Neurosci 1992; 15 (4):133-139.
54. Kubota Y, Mikawa S, Kawaguchi Y. Neostriatal GABAergic interneurons contain NOS, calretinin or parvalbumin. NeuroReport 1993; 5:205-208.
55. Wood JG, Wallace RW, Whitaker JN et al. Immunocytochemical localization of calmodulin and a heat-labile calmodulin-binding protein (CaM-BP₈₀) in basal ganglia of mouse brain. J Cell Biol 1980; 84:66-76.
56. Seto-Ohshima A, Keino H, Kitajima S et al. Developmental change of the immunoreactivity to anti-calmodulin antibody in the mouse brain. Acta Histochem Cytochem 1984; 17:109-117.
57. Biber A, Schmid G, Hempel K. Calmodulin content in specific brain areas. Exp Brain Res 1984; 56:323-326.
58. Teolato S, Calderini G, Bonetti AC et al. Calmodulin content in different brain areas of aging rats. Neurosci Lett 1986; 38:57-60.
59. Hoskins B, Ho JK. Effects of maturation and aging on calmodulin and calmodulin-regulated enzymes in various regions of mouse brain. Mechanisms of Aging and Development 1986; 36:173-186.
60. Berchtold MW, Celio MR, Heizmann CW. Parvalbumin in human brain. J Neurochem 1985; 45:235-239.
61. Kosaka T, Heizmann CW, Tateishi K et al. An aspect of the organizational principle of the gamma-aminobutyric system in the cerebral cortex. Brain Res 1987; 409:403-408.
62. Celio MR. Parvalbumin in most gamma-aminobutyric acid-containing neurons of the rat cerebral cortex. Science 1986; 231:995-997.
63. Demeulemeester H, Vandesande F, Orban GA et al. Heterogeneity of GABAergic cells in cat visual cortex. J Neurosci 1988; 8:988-1000.
64. Demeulemeester H, Arckens L, Vandesande F et al. Calcium binding proteins and neuropeptides as molecular markers of GABAergic interneurons in the cat visual cortex. Exp Brain Res 1991; 84:538-544.
65. Stichel CC, Singer W, Heizmann CW et al. Immunohistochemical localization of calcium-binding proteins, parvalbumin and calbindin-D28k, in the adult and developing visual cortex of cats: a light

and electron microscopic study. J Comp Neurol 1987; 262:563-577.

66. Hendry SHC, Jones EG, Emson PC et al. Two classes of cortical GABA neurons defined by differential calcium binding protein immunoreactivities. Exp Brain Res 1989; 76:467-472.

67. van Brederode JFM, Mulligan KA, Hendrickson AE. Calcium-binding proteins as markers for subpopulations of GABAergic neurons in monkey striate cortex. J Comp Neurol 1990; 298:1-22.

68. DeFelipe J, Hendry SHC, Jones EG. Visualization of chandelier cell axons by parvalbumin immunoreactivity in monkey cerebral cortex. Proc Natl Acad Sci USA 1989; 86:2093-2097.

69. DeFelipe J, Hendry SHC, Jones EG. Synapses of double bouquet cells in monkey cerebral cortex visualized by calbindin immunoreactivity. Brain Res 1989b; 503:49-54.

70. Blümcke I, Hof PR, Morrison JH et al. Distribution of parvalbumin immunoreactivity in the visual cortex of Old World monkeys and humans. J Comp Neurol 1990; 301:417-432.

71. Lewis DA, Lund JS. Heterogeneity of chandelier neurons in monkey neocortex: Corticotropin-releasing factor- and parvalbumin-immunoreactive populations. J Comp Neurol 1990; 293:599-615.

72. Hendrickson AE, van Brederode JFM, Mulligan KA et al. Development of the calcium-binding proteins parvalbumin and calbindin in monkey striate cortex. J Comp Neurol 1991; 307:626-646.

73. Spatz WB, Illing RB, Vogt Weisenhorn DM. Distribution of cytochrome oxidase and parvalbumin in primary visual cortex of the adult and neonate monkey, Callithrix jacchus. J Comp Neurol 1994; 339:519-534.

74. McMullen NT, Smelser CB, De Denecia RK. A quantitative analysis of parvalbumin neurons in rabbit auditory neocortex. J Comp Neurol 1994; 349:493-511.

75. DeFelipe J, Hendry SHC, Hashikawa T et al. A microcolumnar structure of monkey cerebral cortex revealed by immunocytochemical studies of double bouquet cell axons. Neurosci 1990; 37: 655-673.

76. DeFelipe J, Jones EG. High-resolution light and electron microscopic immunocytochemistry of colocalized GABA and calbindin D28K in somata and double bouquet cell axons of monkey somatosensory cortex. Eur J Neurosci 1992; 4:46-60.

77. Alcantara S, Ferrer I, Soriano E. Postnatal development of parvalbumin and calbindin D-28K immunoreactivities in the cerebral cortex of the rat. Anat Embryol 1993; 188:63-73.

78. Hogan D, Berman NEJ. Transient expression of calbindin D28K immunoreactivity in layer V pyramidal neurons during postnatal development of kitten cortical areas. Devel Brain Res 1993; 74:177-192.

79. Frantz GD, Tobin AJ. Cellular distribution of calbindin D28k mRNAs in the adult mouse brain. J Neurosci Res 1994; 37: 287-302.

80. Glezer II, Hof PR, Morgane PJ. Calretinin-immunoreactive neurons in the primary visual cortex of dolphin and human brains. Brain Res 1992; 595:181-188.

81. Glezer II, Hof PR, Leranth C et al. Calcium-binding protein-containing neuronal populations in mammalian visual cortex: a comparative study in whales, insectivores, bats, rodents and primates. Cerebral Cortex 1993; 3:249-272.

82. DeFelipe J. Neocortical neuronal diversity: chemical heterogeneity revealed by colocalization studies of classic neurotransmitters, neuropeptides, calcium-binding proteins and cell surface molecules. Cerebral Cortex 1993; 3:273-289.

83. Hendry SHC, Carder RK. Neurochemical compartmentation of monkey and human visual cortex: Similarities and variations in calbindin immunoreactivity across species. Vis Neurosci 1993; 10:1109-1120.

84. Blümcke I, Hof PR, Morrison JH et al. Parvalbumin in the monkey striate cortex: a quantitative immunoelectron-microscopy study. Brain Res 1991; 554:237-243.

85. DelRio MR, DeFelipe J. A study of SMI 32-stained pyramidal cells, parvalbumin-immunoreactive chandelier cells, and presumptive thalamocortical axons in the human temporal neocortex. J Comp Neurol 1994; 342:389-408.

86. Braun K, Scheich H, Schachner M et al. Distribution of parvalbumin, cytochrome oxidase activity and ^{14}C-2-deoxyglucose uptake in the brain of the zebra finch. I. Auditory and vocal motor systems. Cell Tissue Res 1985; 240:101-115.

87. Braun K, Scheich H, Schachner M et al. Distribution of parvalbumin, cytochrome oxidase activity and ^{14}C-2-deoxyglucose uptake in the brain of the zebra finch. II. Visual system. Cell Tissue Res 1985; 240:117-127.

88. Lewis DA, Snyder CL, Sesack SR. Calretinin-immunoreactive neurons in monkey prefrontal cortex: ultrastructure and associations with dopamine afferents. Soc Neurosci Abstr 1994; 20:578.12.

89. Rogers JH. Immunohistochemical markers in rat cortex: co-localization of calretinin and calbindin-D28k with neuropeptides and GABA. Brain Res 1992; 587:147-157.

90. Daviss SR, Lewis DA. Calbindin- and calretinin-immunoreactive local circuit neurons are increased in density in the prefrontal cortex of schizophrenic subjects. Soc Neurosci Abstr 1993; 19:84.9.

91. Goto S, Singer W, Gu Q. Immunocytochemical localization of calcineurin in the adult and developing primary visual cortex of cats. Exp Brain Res 1993; 96:377-386.

92. Goto S, Nagahiro S, Korematsu K et al. Cellular colocalization of calcium/calmodulin-dependent protein kinase II and calcineurin in the rat cerebral cortex and hippocampus. Neurosci Lett 1993; 149:189-192.

93. Saitoh S, Takamatsu K, Kobayashi M et al. Distribution of hippocalcin mRNA and immunoreactivity in rat brain. Neurosci Lett 1993; 157:107-110.

94. Dyck RH, Van Eldik LJ, Cynader MS. Immunohistochemical localization of the S100ß protein in postnatal cat visual cortex: spatial and temporal patterns of expression in cortical and subcortical glia. Devel Brain Res 1993; 72:181-192.

95. Taylor, AN, Brindak ME. Chick brain calcium-binding protein: comparison with intestinal vitamin D-induced calcium-binding protein. Arch Biochem Biophys 1974; 161:100-108.

96. Jande SS, Maler L, Lawson DEM. Immunohistochemical mapping of vitamin D-dependent calcium-binding protein in brain. Nature 1981; 294:765-767.

97. Jande SS, Tolnal S, Lawson DEM. Immunohistochemical localization of vitamin D-dependent calcium-binding protein in duodenum, kidney, uterus and cerebellum of chickens. Histochem 1981; 71:99-116.

98. Roth J, Baetens D, Norman AW et al. Specific neurons in chick central nervous system stain with an antibody against chick intestinal vitamin D-dependent calcium-binding protein. Brain Res 1981; 222:452-457.

99. Baimbridge KG, Miller JJ. Immunohistochemical localization of calcium-binding protein in the cerebellum, hippocampal formation and olfactory bulb of the rat. Brain Res 1982; 245:223-229.

100. Celio MR, Heizmann CW. Calcium-binding protein parvalbumin as a neuronal marker. Nature 1991; 293:300-302.

101. Braun K, Schachner M, Scheich H, Heizmann CW. Cellular localization of the Ca^{2+}-binding protein parvalbumin in the developing avian cerebellum. Cell Tissue Res 1986; 243:69-78.

102. Stichel CC, Kägi U, Heizmann CW. Parvalbumin in cat brain: isolation, characterization and localization. J Neurochem 1986; 47:46-53.

103. Lin CT, Dedman JR, Brinkley BR et al. Localization of calmodulin in rat cerebellum by immunoelectron microscopy. J Cell Biol 1980; 85:473-480.

104. Caceres A, Bender P, Snavely L et al. Distribution and subcellular localization of calmodulin in adult and developing brain tissue. Neurosci 1983; 10:449-461.

105. Seto-Ohshima A, Sano M, Mizutani A. Characteristic localization of calmodulin in human tissues: immunohistochemical study in the paraffin sections. Acta Histochem Cytochem 1985; 18:275-282.

106. Goto S, Matsukado Y, Uemura S et al. A comparative immunohistochemical study of calcineurin and S-100 protein in mammalian and avian brains. Exp Brain Res 1988; 69:645-650.

107. Rogers JH. Calretinin: a gene for novel calcium-binding protein expressed principally in neurons. J Cell Biol 1987; 105:1343-1353.

108. Sequier J-M, Hunziker W, Richards G. Localization of calbindin D-28KmRNA in rat tissues by in situ hybridization. Neurosci Lett 1988; 86:155-160.

109. Pasteels B, Pochet R, Surardt L et al. Ultrastrucural localization of brain "Vitamin D-dependent" calcium binding proteins. Brain Res 1986; 384:294-303.

110. Kadowaki K, McGowan E, Mock G et al. Distribution of calcium binding protein mRNAs in rat cerebellar cortex. Neurosic Lett 1993; 153:80-84.

111. Pinol HR, Kägi U, Heizmann CW et al. Poly- and monoclonal antibodies against recombinant rat brain calbindin D-28K were produced to map its selective distribution in the central nervous system. J Neurochem 1990; 54:1827-1833.

112. Scotti AL, Nitsch C. Differential Ca²+ binding properties in the human cerebellar cortex: distribution of parvalbumin and calbindin D-28k immunoreactivity. Anat Embryol 1992; 185:163-167.

113. Katsekos CD, Frankfurter A, Christakos S et al. Differential localization of class III - tubulin isotype and calbindin-D28K defines distinct neuronal types in the developing human cerebellar cortex. J Neuropathol and Exp Neurol 1993; 52:655-666.

114. Plogmann D, Celio MR. Intracellular concentration of parvalbumin in nerve cells. Brain Res 1993; 600:273-279.

115. Kosaka T, Kosaka K, Nakayama T et al. Axons and axon terminals of cerebellar Purkinje cells and basket cells have higher levels of parvalbumin immunoreactivity than somata and dendrites: quantitative analysis by immunogold labeling. Exp Brain Res 1993; 93:483-491.

116. Tolosa de Talamoni N, Smith CA, Wasserman RH et al. Immunocytochemical localization of the plasma membrane calcium pump, calbindin D-28K and parvalbumin in Purkinje cells of avian and mammalian cerebellum. Proc Natl Acad Sci USA 1993; 90:11949-11953.

117. Brorson JR, Bleakman D, Gibbons SJ et al. The properties of intracellular calcium stores in cultured rat cerebellar neurons. J Neurosci 1991; 11:4024-4043.

118. Rogers JH. Calretinin. In: Heizmann CW, ed. Novel Calcium Binding Proteins. Berlin: Springer Verlag, 1991:251-276.

119. Arai R, Winsky L, Arai M et al. Immunohistochemical localization of calretinin in the rat hindbrain. J Comp Neurol 1991; 310:21-44.

120. Arai R, Jacobowitz DM, Deura S. Ultrastructural localization of calretinin immunoreactivity in lobule V of the rat cerebellum. Brain Res 1993; 613:300-304.

121. Rogers JH. Immunoreactivity for calretinin and other calcium binding proteins in the cerebellum. Neuroscience 1989; 31:711-721.

122. Resibois A, Rogers JH. Calretinin in rat brain: an immunohistochemical study. Neuroscience 1992; 46:101-134.

123. Braak E, Braak H. The new monodendritic neuronal type within the adult human cerebellar granule cell layer shows calretinin-immunoreactivity. Neurosci Lett 1993; 154:199-202.
124. Yamakuni T, Usui H, Iwanaga T et al. Isolation and immunohistochemical localization of a cerebellar protein. Neurosci Lett 1984; 45:235-240.
125. Yamakuni T, Araki K, Takahashi Y. The developmental changes of mRNA levels for a cerebellar protein (spot 35 protein) in rat brains. FEBS Lett 1985; 188:127-130.
126. Dawson TM, Steiner JP, Lyons WE et al. The immunophilins, FK506 binding protein and cyclophilin, are discretely localized in the brain: relationship to calcineurin. Neurosci 1994; 62:569-580.
127. Lenz SE, Henschel Y, Zopf D et al. VILIP, a cognate protein of the retinal calcium binding proteins visinin and recoverin, is expressed in the developing chicken brain. Molec Brain Res 1992; 15:133-140.
128. Ito A, Hashimoto T, Hirai M et al. The complete primary structure of calcineurin A, a calmodulin binding protein homologous with protein phosphatase A and 2A. Biochem Biophys Res Commun 1989; 163:1492-1497.
129. Goto S, Matsukado Y, Miyamoto E et al. Morphological characterization of the rat striatal neurons expressing calcineurin immunoreactivity. Neurosci 1987; 22:189-201.
130. Zuschratter W, Scheich H, Heizmann CW. Ultrastructural localization of the calcium-binding protein parvalbumin in neurons of the song system of the zebra finch. Cell Tissue Res 1985; 241:77-83.
131. Celio MR, Keller GA, Bloom FA. Immunoelectron microscopy of neural antigens on ultrathin frozen sections. J Histochem Cytochem 1986; 34:491-500.
132. Legrand C, Thomasset M, Parkes CO et al. Calcium binding protein in the developing rat cerebellum. Cell Tissue Res 1983; 233:389-402.
133. Aronin N, Chase K, Folsom R et al. Immunoreactive calcium-binding protein (calbindin D28K) in interneurons and trigeminothalamic neurons of the rat nucleus caudalis localized with peroxidase and immunogold methods. Synapse 1991; 7:106-113.
134. Legrand C, Clos J, Legrand J et al. Localization of S100 protein in the rat cerebellum: an immunoelectron microscopy study. Neuropathol Appl Neurobiol 1981; 7:299-306.
135. Llinas R, Hess R. The role of calcium in neuronal function. In: Schmitt FO, Worden FG, eds. The Neurosciences: Fourth Study Program. Cambridge Massachussetts-London: MIT Press, 1979: 555-571.
136. Lynch G, Baudry M. The biochemistry of memory: a new and specific hypothesis. Science 1984; 224:1057-1063.
137. Fifkova E. Actin in the nervous system. Brain Res Rev 1985; 9:187-215.

CHAPTER 5

BIOLOGICAL FUNCTIONS OF EF-HAND CA^{2+}-BINDING PROTEINS

MOLECULAR INVESTIGATIONS

The consensus amino acid sequence for an EF-hand Ca^{2+}-binding domain has allowed identification of more than 200 Ca^{2+}-binding proteins from primary structures, many of them expressed in the central nervous system (Table 2.3). A central question concerns the biological function of these proteins. Ca^{2+}-binding proteins with known functions, such as calmodulin, troponin C, myosin-light chains, calpain or calcineurin, are far outnumbered by those whose roles are not known. Most of them are expressed in a cell-type specific manner (in contrast to calmodulin, which is ubiquitously distributed). Recently, this family of proteins has attracted a lot of interest since altered concentrations of some of them have been reported in several disease states of the central nervous system and in tumor cells (chapter 6). An exploration of their involvement in mechanisms of calcium-mediated regulation in normal cells should therefore help to elucidate the underlying mechanisms of these pathological conditions.

Attempts to identify functional roles have been quite successful in the case of calmodulin, as many probable calmodulin acceptor proteins have been identified in the cytosol[1] and nucleus.(Table 1.1)

Suggested functions of some of the other Ca^{2+}-binding proteins are listed in Table 2.3 and will not be discussed in more detail. Gene transfers, or even better, gene disruption experiments are thought to be useful tools for functional studies of EF-hand proteins that are expressed in a tissue-specific fashion. Over-expression or disruption of the genes for these proteins is not expected to interfere with the basic cellular metabolism, but rather with a specialized function of the respective cell type. This approach should yield interesting information on some of the distinct functions and targets of this large family of proteins.

Recent progress has been made in the elucidation of the physiological roles of the cytosolic Ca^{2+}-binding proteins, parvalbumin and calbindin D-28K, and of the S100 proteins, and will therefore be discussed here in more detail.

A) CYTOSOLIC CA²⁺-BINDING PROTEINS, PARVALBUMIN, AND CALBINDIN D-28K

Parvalbumin[2] is present in a subpopulation of mammalian neurons containing the inhibitory neurotransmitter γ-aminobutyric acid (GABA) (chapter 4). Parvalbumin is generally associated with neurons that have a high firing rate and a high oxidative metabolism. The presence of parvalbumin in fast-spiking cells (firing at high frequency and showing no adaptation of spike frequency with sustained depolarization) in the rat hippocampus (CA2 region) has been demonstrated by injection of lucifer yellow in vitro in combination with post-embedding parvalbumin immunohistochemistry. It is suggested that parvalbumin may be involved in the buffering/transport of calcium in a subset of neurons with specialized electrophysiological properties.

Calbindin D-28K[3] is present in a subpopulation of neurons scattered in most but not all areas of the CNS. The role of this protein in neurons is also suggested to be that of a Ca^{2+}-buffering transport protein.

It was assumed that neurons containing parvalbumin or calbindin would be more resistant to the cellular degeneration observed in epilepsy and ischemia because of their greater capacity to buffer Ca^{2+} and protecting cells from Ca^{2+} overload. However, investigations of the vulnerability of such neurons in the human brain as well as in experimental animal models have revealed contradictory results,[4] mainly because they were solely based on immunohistochemical localization studies, and not on any direct functional and quantitative approach.

In order to get direct evidence for the hypothesis that the Ca^{2+} buffering effect of parvalbumin shortens repolarization time, human neuroblastoma cells (SKNBE cells containing no endogenous parvalbumin) were made to ectopically express parvalbumin and calcium fluxes were monitored with Fura-2/AM under stimulation conditions. The procedure is illustrated in Figure 5.1.

SKNBE cells were transfected with the parvalbumin gene as outlined in Figures 5.1 and 5.2, and a representative sample was stained with an antiserum against parvalbumin. Immunostaining and quantitative Western blot analysis confirmed that this resulted in physiological levels of parvalbumin (~0.5 mM within a cell) in about 10-20% of the transfected cells.

The SKNBE cells were depolarized with KCl and changes in intracellular calcium levels were monitored with Fura 2-/AM (Fig. 5.2). No rise of Ca^{2+} upon stimulation was observed in parvalbumin-expressing cells, indicating that parvalbumin can impair the Ca^{2+} response in stimulated excitable cells. These results then support the hypothesis that parvalbumin can act as a Ca^{2+} buffer and can play an excito-protective role against Ca^{2+} overload in neurons.

These results are in good agreement with another stable transfection study (calbindin D-28K), using the pituitary tumor cell line GH$_3$ and demonstrating an alteration of calcium currents and intracellular calcium homeostasis.[5]

In a different set of experiments parvalbumin has been quantified within Purkinje cells of the cerebellum by electron microscopy.[6] Parvalbumin was found to be inhomogeneously distributed within these cells. The axons and axon terminals had significantly higher levels of parvalbumin-immunoreactivity (1 mM parvalbumin or more) than somata, dendrites and dendritic spines (50-100 μM). A similar result was reported for calbindin D-28K in a WEHI7.2 thymoma cell line.[7]

This differential distribution of parvalbumin in neuronal compartments suggests that this protein might be a mobile buffer/Ca^{2+} shuttle within neurons, as suggested for skeletal muscles[8-10] and might be discussed in relation to the spatiotemporal aspects of Ca^{2+}

Parvalbumin Transfection and Ca2+ Measuring in Human Neuroblastoma Cells

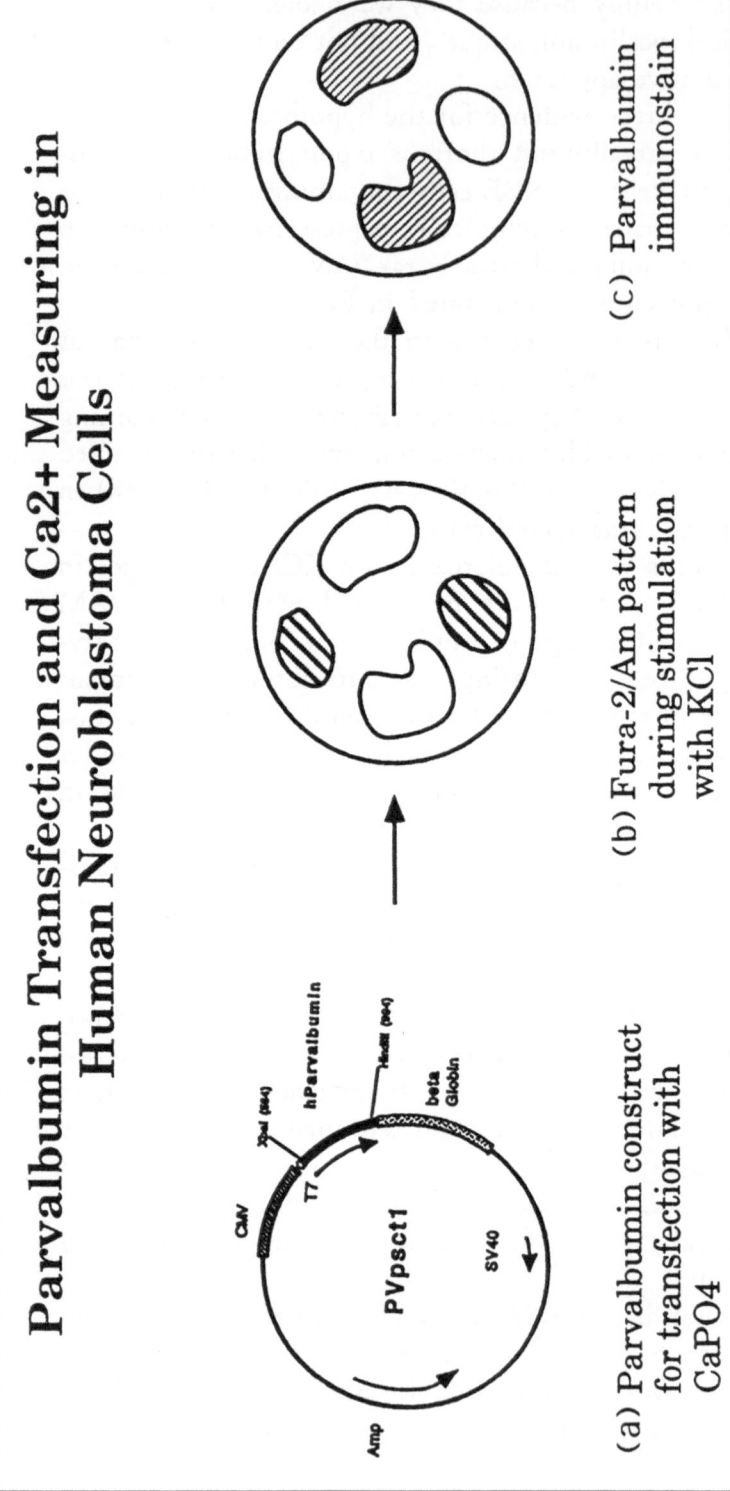

(a) Parvalbumin construct for transfection with CaPO4

(b) Fura-2/Am pattern during stimulation with KCl

(c) Parvalbumin immunostain

Fig. 5.1. Schematic presentation of the transient transfection of human parvalbumin into human neuroblastoma (SKNBE) cells.
(a) plasmid construct; full length parvalbumin cDNA is directly cloned into the eukaryotic expression vector, pSCT1, which has a strong CMV promoter for high permanent expression.
(b) Imaging of $[Ca^{2+}]_i$ by Fura-2-AM to identify non-transfected cells responding to depolarization with an elevation of $[Ca^{2+}]_i$ (barred cells) from parvalbumin-transfected cells (white cells) with no change of $[Ca^{2+}]_i$.
(c) Immunostaining. After depolarization, cells were stained for parvalbumin. Parvalbumin expressing cells (shaded cells) are expected to show no increase of $[Ca^{2+}]_i$ upon depolarization.
Reprinted from: Lutum C, Dreessen J, Knöpfel T et al. Ca^{2+} response in parvalbumin transfected cells. Third European Symposium on Ca^{2+}-binding Proteins in Normal and Transformed Cells. Zurich: 1994;81 (abstract); manuscript in preparation.

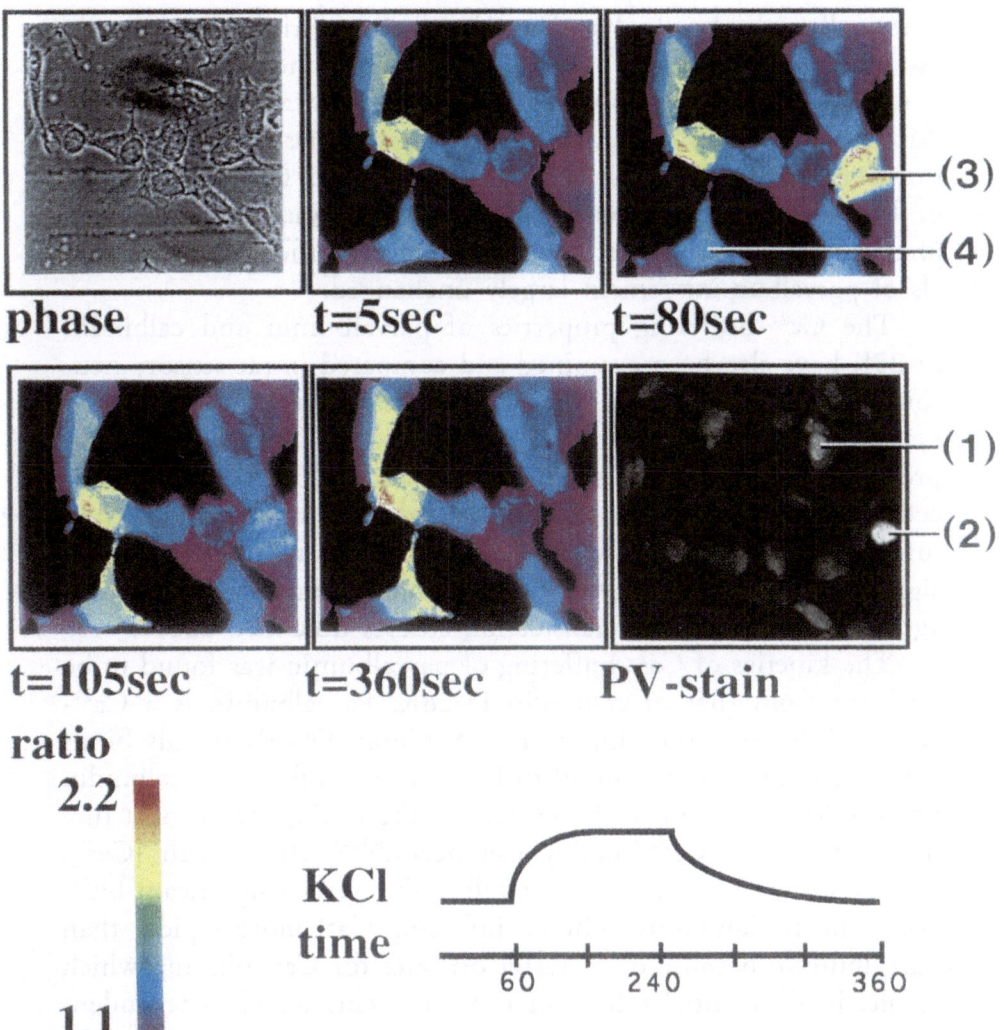

Fig. 5.2. SKNBE cells were transiently transfected with full-length human parvalbumin cDNA and stimulated with KCl, resulting in a Ca²⁺ increase only in those cells lacking parvalbumin.
Phase contrast pattern (top, left); followed by Fura-2-AM imaging in the ratio 340/390 nm emmission at 505 nm (red color shows high [Ca²⁺]$_i$ levels) after various times of KCl stimulation; immunostain pattern with an antibody against parvalbumin (bottom, right).
Only cells (1) and (2) show clear parvalbumin staining; these cells do not show an increase of [Ca²⁺]$_i$ upon stimulation, however, cells (3) and (4) containing no parvalbumin show a strong increase in [Ca²⁺]$_i$.
Reprinted from: Lutum C, Dreessen J, Knöpfel T et al. Ca²⁺ response in parvalbumin transfected cells. Third European Symposium on Ca²⁺-binding Proteins in Normal and Transformed Cells. Zurich: 1994;81 (abstract); manuscript in preparation.

signaling[11] where local Ca^{2+} buffering and transport is needed. The spatial distribution of parvalbumin in Purkinje cells[6] also correlates well with the results of a microfluorometric imaging study on cerebellar Purkinje cells.[12] In this study, the spatiotemporal dynamics of intracellular calcium concentration are demonstrated during spontaneous and evoked activity and it has been found that intradendritic $[Ca^{2+}]_i$ increases 10- to 100-fold (possibly because of low levels of parvalbumins) between the inactive and active phases of oscillation, whereas somatic levels (buffered by high levels of parvalbumin) remain largely unchanged.

The Ca^{2+} buffering properties of parvalbumin and calbindin D-28K have also been examined and compared in rat sensory neurons after brief depolarizations of rat dorsal root ganglion (DRG) neurons[13] and in other cells.[14,15] Introduction of both Ca^{2+}-binding proteins into neurons was achieved by adding them to the intracellular solution in the patch pipette and allowing them to diffuse into the neuron in the whole cell configuration.[13] Both proteins significantly affected depolarization-induced changes in $[Ca^{2+}]_i$ in agreement with the cell transfection studies described above.

The kinetics of Ca^{2+} buffering of parvalbumin was found to be different from that of calbindin D-28K. Parvalbumin is a Ca^{2+}- and Mg^{2+}-binding protein, whereas calbindin D-28K mainly binds Ca^{2+}. Parvalbumin was found to be a slower buffer than calbindin because Mg^{2+} (bound to the protein in the resting state) must first dissociate before Ca^{2+}-binding can occur.[8,16-20] As a result $[Ca^{2+}]_i$ increases more rapidly before parvalbumin exerts a significant buffering effect. Calbindin buffered influxing Ca^{2+} more rapidly than parvalbumin, because of its faster on-rates for Ca^{2+}-binding which results in differential effects of those proteins on Ca^{2+}-dependent processes in neurons.

All these observations show that parvalbumin and calbindin are effective Ca^{2+} buffers in neurons and probably operate over different time scales. This buffer function provides a mechanism for protecting neurons against Ca^{2+}-mediated excitotoxicity.

At present, information on new Ca^{2+}-binding proteins in the brain, their primary structures and the regulation of their expression is accumulating at a rapid pace (see Table 3.1). The most difficult and interesting question to date remains that of their precise biological function and their mechanisms of action and interaction.

B) INTRA- AND EXTRACELLULAR FUNCTIONS OF S100 PROTEINS

The first description of S100 proteins was made by Moore who initially characterized a group of low molecular weight (10-12 kD) acidic proteins that were enriched in the nervous system.[21] Subsequently, the S100A1 (S100α) and S100β species were extracted from brain and identified, and over the past few years a number of other S100 proteins have been described that are expressed in a cell-type specific fashion.[22-25] The clustered organization of most S100 genes on human chromosome 1q21 has permitted to introduce a new logical nomenclature for these genes based on their physical arrangement on this chromosome (Table 2.2, and Fig. 3.1). S100 proteins appear to be involved in the regulation of a number of cellular processes from cell cycle progression to differentiation and tumor progression as well as extracellular functions. S100 proteins probably exert their biological effects by interacting with target proteins (e.g. annexins) in a Ca²⁺-dependent fashion similar to calmodulin.

S100 proteins may exist as monomers, dimers, or even oligomers, and it is suggested that the dimeric or oligomeric forms are the active extracellular protein species. Furthermore, S100β,[26] S100A6 (calcyclin),[27] and probably other S100 proteins also bind Zn²⁺ with high affinities.

Some progress has been made in the elucidation of the extra- and intracellular roles of S100β expressed in glial cells of the CNS. The gene for S100β has been assigned to chromosome 21q22[28] (Table 2.2), suggesting a causal relationship to the defects of the nervous system in patients with trisomy 21 (Down's syndrome). S100β protein expression is increased in the astrocytes and microglia of patients with Down's syndrome and Alzheimer's disease. Remarkably, the neuropathology of both disorders is very similar. For example, abnormal phosphorylation of MAP proteins (including tau) is involved in the constitution of neurofibrillary tangles. S100β is overexpressed in both diseases and was found to modulate tau phosphorylation, suggesting that it may play an important role in these disorders.[29] This was supported by other studies[30] and by the observation that the brains of transgenic mice bearing multiple copies of the S100β gene show changes comparable to those in Down's syndrome.[31,32]

These experimental approaches may provide the means for better understanding of the cellular and molecular basis for the mental retardation in these diseases. However, this issue is still controversial. For example, in a study in which the human S100β transgene was expressed at a high level in a copy number-dependent and cell-specific manner in transgenic mice,[32] these mice did not manifest any obvious behavioral differences or pathological abnormalities at autopsy when compared to their normal littermates, even though the increased S100β immunoreactivity of astrocytes in their brains was similar to that observed in the brains of Down's syndrome and Alzheimer patients. This implied that 10- to 100-fold overexpression of the S100β gene was well tolerated. However, subtle changes were not excluded in this study. In another study of transgenic mice carrying multiple copies of the human S100β gene[33] a behavioral abnormality was seen (hyperactivity in females but not in males).

A further study demonstrated[34] that in the Lewis rat, autoimmune T cells specific for the astrocyte-derived autoantigen, S100β, adoptively transfer an unusual type of inflammatory CNS disease, described as experimental autoimmune panencephalitis. The clinical characteristics and the histopathology of this new model of tissue-specific autoimmunity closely resemble the human disease, multiple sclerosis.

Like several other Ca^{2+}-binding proteins, S100β is also secreted from cells (Table 1.2) and may have extracellular functions.[35,36]

The dimeric form of S100β was shown to promote/stimulate neurite extension in chick embryonic cortical neurons and murine neuroblastoma cells.[37,38] These extracellular effects of S100β were also demonstrated in vivo.[31,39] Both astrocytosis and neurite proliferation occur in transgenic mice, which express high levels of S100β, demonstrating that S100β is a growth factor in the brain and that this animal model may be useful for studying the role of S100β in glial-neuronal interactions and in normal and diseased brains.

It has also been shown[40] that incubation of PC12 cells with S100β induces a rapid rise of intracellular Ca^{2+} followed by an increase in cells undergoing degeneration and apoptosis. The mechanism of action is unknown. A search for cell surface receptors for S100 proteins involved in signal transduction is presently underway.

The mechanism of secretion is completely unknown since S100 proteins have no classical leader peptide. One possibility is that

Ca^{2+} binding results in an exposure of a hydrophobic domain at the N-terminus of S100 proteins which is then able to interact with the membrane or membranous proteins (e.g. annexins) resulting in secretion of the molecule.

Several S100 proteins are differentially regulated in tumor cells, suggesting an involvement in tumorigenicity. For example, S100A4 has been isolated by subtractive hybridization of non-metastatic cells from metastatic rat and mouse mammary cells.[41,42] More recently, transfection experiments showed that S100A4 can induce a metastatic phenotype in rat benign mammary epithelial cells.[43] The expression of this protein has been found to be elevated in serum-stimulated mouse fibroblasts,[44] in src-transformed rat fibroblasts,[45] in murine fibroblasts transformed either by oncogenes or by chemical carcinogens, in metastatic mouse-cell lines[42] and in v-K-ras-transformed rat kidney cells.[46] S100A4 expression may be inversely regulated to the putative metastasis suppressor gene *nm23*.[47]

S100A6 was initially identified as an mRNA preferentially expressed in proliferating rather than quiescent cells.[44,48] Increased expression levels of S100A6 have been correlated to growth of hair follicles,[49] to transformation and metastasis of NIH-3T3 cells[50] and to the metastasizing behavior of human melanoma cells.[51]

In contrast, S100A2 has been isolated as a clone specific for normal human mammary epithelial cells and seems to be down-regulated in tumorigenic cells.[52] The human chromosomal region 1q21,[53,54] where most of the S100 protein genes are tightly clustered, is a region known to have lower stability and frequent loss of heterozygosity in breast cancer.[55,56,59] In order to understand how the distinct S100 proteins are regulated during tumorigenesis and in the onset of metastatic phenotype of human mammary epithelial cells, a panel of human breast-cancer cell lines and tissues was screened for expression of S100A4, S100A6, S100A2, and S100A1 and S100β, and the results were compared with a variety of molecular prognostic factors.[58] Each S100 protein is shown to be individually regulated in the human breast-cancer cell lines, but it appears that the expression levels of S100 proteins do not strictly correlate with prognostic factors or the tumorigenicity of the cells.

There is, however, a significant correlation between enhanced expression of S100A4 and presence of the invasivity marker urokinase-type plasminogen activator (uPA).[58] This observation supports the idea that S100A4 may play an important role in the acquisition of metastatic potential by human mammary epithelial cells, although the mechanism of action is not yet known. Recently, the non-muscle myosin heavy chain has been identified as a possible target for S100A4,[55] which might influence the metastatic properties of tumor cells by altering cell motility.[57] Clearly, there is a link between the expression of some S100 proteins and the metastasizing behavior of tumor cells, and this merits further investigation, especially on tumors of the brain.

PHYSIOLOGICAL AND ANATOMICAL INVESTIGATIONS

RELATIONSHIP BETWEEN CA²⁺-BINDING PROTEINS AND TRANSMITTER SYSTEMS

In most brain regions investigated so far, the parvalbumin-, calbindin D-28K- and calretinin-immunoreactive neurons represent different subpopulations of GABAergic interneurons. In the hippocampal formation, almost all parvalbumin-immunoreactive neurons, which are concentrated in the granule cell layer and hilus of the dentate gyrus and in the stratum pyramidale and stratum oriens of the CA1 and CA3 region, were also found to be immunoreactive with antiserum raised against glutamic acid decarboxylase (GAD). Depending on the layer, between 1% and 50% of GAD-positive neurons also contained parvalbumin.[60] Similar relationships hold true for the somatosensory cortex, where about 70% of all GABA-immunoreactive neurons contain parvalbumin;[61] in rat visual cortex;[62-64] in the dorsal lateral geniculate nucleus of cats[65] and in the external plexiform layer of the olfactory bulb.[66] In the striatum the parvalbumin-immunoreactive neurons represent GABAergic interneurons, while calbindin D-28K-immunoreactive neurons represent GABA/glutamatergic striatonigral projection neurons.[67] In the neocortex of *Macaca fasciculata* most parvalbumin- and calbindin D-28K-positive cells also display GABA-immunoreactivity and a relatively small number of GABA-immunoreactive cells lack immunoreactivity for both Ca²⁺-binding proteins.[68] In area 17 of *Macaca fasciculata*, calbindin D-28K- and parvalbumin-immunostaining is found in separate populations of GABA-positive neurons and afferent axons. Calbindin D-28K-immunoreactive neurons make up a large proportion of the GABA-positive cells in layers II and III. Parvalbumin-immunoreactive somata and neuropil are present in all layers with highest densities in layers IVA and IVC. However, most of the parvalbumin-positive processes in layers II and III and in layers IVA and IVC do not display GABA-immunoreactivity. Other examples of the colocalization of parvalbumin- and GABA- and/or GAD-immunoreactivity are the cerebellar Purkinje cells (see above).

The majority of calretinin-immunoreactive interneurons in the hippocampal formation is GABAergic with the exception of one morphologically distinct subpopulation of calretinin-immunoreactive neurons, the spiny neurons, which are not GABA-immunoreactive.[69]

On the other hand, there are several brain systems where a relationship between immunoreactivity against parvalbumin, calbindin D-28K or calretinin and GABAergic mechanisms is not obvious, e.g. the GABAergic cerebellar Golgi type II cells, which do not exhibit parvalbumin-immunoreactivity or, conversely, in the parvalbumin-immunoreactive ganglion cells of the cat retina, which are not thought to be GABAergic.[70] The calretinin-immunoreactive fibers terminating in the superficial region of the granule cell layer in the dentate gyrus represent excitatory substance P containing afferents from the hypothalamus.[71]

Furthermore, other neurotransmitters can be present in parvalbumin-, calbindin D-28K-and calretinin-positive neurons.[72] For instance, in rat brain a colocalization of calbindin D-28K- and tyrosine hydroxylase (TH)-immunoreactivity has been found in the ventral tegmental area, in the dorsal tier of the substantia nigra pars compacta and in the retro-rubral area.[72] In these systems, calbindin D-28K-immunoreactive neurons form a subpopulation of dopaminergic neurons projecting to the striatal matrix. The distribution of the non-calbindin D-28K-immunoreactive dopaminergic neurons that lie in the ventral tier of the pars compacta and in the pars reticulata matches the origin of the dopaminergic projection to the striatal patches.

These examples show that the interrelationships between the neuronal subpopulations characterized by either Ca²⁺-binding protein and GABA or GAD do not follow a common rule but show characteristic features for each brain area. Thus, parvalbumin-, calretinin- and calbindin D-28K-immunoreactivity of neuron populations might rather reflect distinct electrophysiological properties of neurons such as firing patterns or electrical and/or metabolical activity levels independently of the transmitter they use (see below).

Under the assumption that neurodegenerative processes are mediated by excitotoxic amino acids, the differential sensitivity of neurons containing calcium-binding proteins to these transmitters is of interest. For parvalbumin-immunoreactive neurons a selective vulnerability to the non-NMDA (n-methyl-d-aspartate) agonists, kainate and AMPA, but not to NMDA, was reported in vitro[73] as well as in vivo in striatal parvalbumin-positive neurons.[74] In contrast, other groups found an enhancement of NMDA neurotoxicity in postmitotic cultivated cortical neurons after they had been transfected with the parvalbumin gene.[75] In addition, these authors found that the expression of parvalbumin in these neurons enhances the release of aspartate and glutamate. Calbindin D-28K-immunoreactive cultured hippocampal neurons seem to be relatively resistant to glutamate-induced neurotoxicity[76,77] and calretinin-immunoreactive neurons tolerate toxic concentrations of glutamate, NMDA, kainate and quisqualate in vitro.[78] However, this latter finding is in marked contrast to in vivo observations that calretinin-immunoreactive neurons are among the most vulnerable neurons in neurodegenerative diseases and in animal models (see chapter 6).

Almost all parvalbumin-immunoreactive GABAergic interneurons in CA1, CA3 and in the dentate gyrus express muscarinergic receptors, which is indicative for a cholinergic input from the medial septum and diagonal band.[79,80]

Calbindin D-28K-levels rapidly increase in cerebellar slices after they have been exposed to excitatory and excitotoxic concentrations of glutamate or its analogue kainic acid.[81] This increase is reversible, can be blocked by CNQX and is independent of Ca^{2+} influx. This strongly supports the idea that Purkinje cells can increase their Ca^{2+} buffering capacity when their kainate/AMPA receptors are activated. Interestingly, Purkinje cells seem to have different thresholds for maximally activating their calbindin D-28K but it is unclear if the increase of calbindin D-28K is due to de novo synthesis. This excitation-induced increase of calbindin D-28K-immunoreactivity is in accordance with the rise in calbindin D-28K mRNA after perforant path stimulation, which is glutamatergic,[82] but it is somewhat in contrast to earlier findings in hippocampal granule cells. Calbindin D-28K-immunoreactivity in dentate granule cells and their dendrites is reduced to about 33% after commissural kindling stimulation, which is also mediated by glutamate.[83-85]

A conclusive interpretation of the various effects summarized above is not yet possible and the interrelation between neurotransmission and the regulation of intracellular Ca²⁺-binding proteins has to be further analyzed.

RELATIONSHIP TO NEURONAL ACTIVITY PATTERNS AND SYNAPTIC PLASTICITY

Table 5.1 summarizes some experimental data demonstrating activation-induced changes of intracellular levels of various Ca²⁺-binding proteins.

The results from a variety of neuronal stimulation/deprivation experiments are puzzling and contradictory and again, a conclusive interpretation of these data is not possible. In the kindling model, stimulation of the perforant path leads to a reduction only of parvalbumin-immunoreactive dentate basket cells,[86,87] but if kindling stimulation is applied, e.g. in the commissural fibers, the number of parvalbumin-positive neurons increases.[88] Two explanations are offered for this phenomenon by the latter authors. Either the increase in cell density is the result of a kindling-induced increase of the parvalbumin content in the cell bodies not identified in controls as a result of a below-threshold level of parvalbumin; or this increase of immunoreactivity may be the result of a kindling-induced shift in the balance between the Ca²⁺-loaded and -unloaded form of parvalbumin towards the former as a result of increased intracellular Ca²⁺-concentrations occurring after seizure activity.[89,90] Parvalbumin recognized on Western blotting by this antiserum seems to be the preferred Ca²⁺-loaded form of parvalbumin.[91]

As already mentioned in the preceding section, stimulation of the perforant path leads to a transient increase of calbindin D-28K-mRNA,[82] whereas amygdaloid or commissural kindling leaves calbindin D-28K-mRNA levels unchanged, but calbindin D-28K-immunoreactivity of dentate granule cells is decreased.[85,92] Biochemical analysis revealed that the immunocytochemically detected loss of calbindin D-28K is due to a decrease in the protein itself. The consequence of this loss of calbindin D-28K in granule cells may be a significant alteration of Ca²⁺-dependent processes underlying

Table 5.1. Activity-dependent regulation of Ca^{2+}-binding proteins

protein	manipulation	area	species	change	structure	references
Parvalbumin	kindling	hippocampus	rat	↑	neuropil	1
	20´ischaemia	hippocampus	gerbil	↑	somata, neuropil (15 min)	2
	20´ischaemia	hippocampus	gerbil	↓ ↑	CA1 (1h), CA3+ hilus (6h),CA1(1-2d)	2
	20´ ischaemia	hippocampus	rat	↓	CA1,/CA3c, hilus (4d)	3
	kainate injection	hippocampus	rat	↓	neurons (7d)	4,5
	stroke	hippocampus	rat	↓	CA1	6
	monocular deprivation (sensitive phase)	visual cortex (binocular part)	rat	↓	neurons, process. (16d)	7
	unilateral dissection	visual system	rat	↑ ↓	neurons terminals	8
	monocular enucleation	lateral geniculate	monkey	↓	neuropil	9
	monocular enucleation	lateral geniculate,	monkey	↓↓	neuropil	10
		striate cortex		↓	neuropil	10
Calbindin D-28K	kindling	dentate gyrus	rat	↓	granule cells	11,12
	PP stimulation	dentate gyrus	rat	↑	CABPmRNA in granule cells	13,14
	electroconvulsive shocks	dentate gyrus	rat	↓	granule cells and mossy fibers	15
	20´ ischaemia	hippocampus	rat	↓	CA1	3
	10´ ischaemia	hippocampus	rat	↑ ↓	CaBPmRNA	13,14
	social deprivation	caudate/putamen	monkey	↓↓	axons, terminals	16
	unilateral dissection	visual system	rat	↓	neurons	8
	monocular enucleation	lateral geniculate,	monkey	↓↓	neuropil	10
		striate cortex		↓	neuropil	10
Calretinin	unilateral TTX- blockade or cochlea extraction	n. magnocellularis n. laminaris	chick	↓	axons (NM), dendrites (NL)	17
	20´ischaemia	hippocampus	rat	↓	hilus, CA3 (12-24h) CA1, CA3, dg (1-3d)	18
	kainate injection	hippocampus	rat	↓	hilus, CA3	19
	eye enucleation	superior colliculus (contralateral)	rat	↑	number of CaR-positive neurons	20

References
1. Kamphuis W, Huisman E, Wadman WJ et al. Brain Res 1989; 479:23-34.
2. Tortosa A, Ferrer I. Neuroscience 1993; 1:33-43.
3. Johansen FF, Tonder N, Zimmer J et al. Neurosci Lett 1990; 120:171-174.
4. Best N, Mitchell J, Baimbridge KG et al. Neurosci Lett 1993; 155:1-6.
5. Best N, Mitchell J, Wheal HV. Acta Neuropathol 1994; 87:187-195.
6. De Jong GI, van der Zee EA, Bohus B et al. Stroke 1993; 24:2082-2086.
7. Cellerino A, Siciliano R, Domenici L et al. Neuroscience 1992; 51:749-753.
8. Schmidt-Kastner R, Meller D, Eysel UT. Exp Neurol. 1992; 17:230-246.
9. Tigges M, Tigges J. Visual Neuroscience 1993; 10:1043-1053
10. Blümcke I, Weruaga E, Kasas S et al. Visual Neuroscience 1994; 11:1-11.
11. Baimbridge KG, Miller JJ. Brain Res 1984; 324:85-90.
12. Baimbridge KG, Mody I, Miller JJ. Epilepsia 1985; 26: 460-465.
13. Lowenstein DH, Miles MF, Hatam F et al. Neuron 1991; 6:627-633.
14. Lowenstein DH, Gwinn RP, Seren S et al. Molec Brain Res 1994; 22:299-308.
15. Tonder N, Kragh J, Bolwig T et al. Hippocampus 1994; 4: 79-83.
16. Martin LJ, Spicer DM, Lewis MH, et al. J Neurosci 1991; 11:3344-3358.
17. Braun K, Rogers JH, Rubel EW. 13th Midwinter Research Meeting ARO, St Petersburg Beach, FL 1990; 388-389.
18. Freund TF, Magloczky Z. Neuroscience 1993; 56:581-596.
19. Magloczky Z, Freund TF. Neuroscience 1993; 56:317-336.
20. Arai M, Arai R, Sasamoto K et al. Brain Res 1993; 613: 341-346.

synaptic transmission and cellular activity in this region. Decreased Ca^{2+}-buffering could lead to augmented transmitter release in the mossy fiber terminals, which in turn results in an increased excitability on the part of the CA3 target neurons. On the other hand, the decreased ability on the part of the Ca^{2+}-buffering could enhance Ca^{2+}-spike generation and the subsequent inhibitory phase resulting from a Ca^{2+}-dependent k^{7}-efflux.[93]

The involvement of calmodulin and its regulated reactions in kindling mechanisms has been biochemically evaluated.[94] After septal kindling, the Ca^{2+}/calmodulin-dependent phosphorylation of synaptic plasma membrane proteins is markedly reduced in the hippocampus and in the amygdaloid-entorhinal complex and lasts for eight weeks post stimulus.

A correlation between Ca^{2+}-binding protein content and firing properties of different non-pyramidal cell types is apparent.[95-99] The parvalbumin-containing basket and axo-axonic cells are fast-firing, whereas neurons arborizing in the dendritic layers show different firing patterns. The striatal GABAergic parvalbumin-immunoreactive neurons belong to a class of fast-spiking interneurons,[100] i.e. they exhibit quite similar physiological characteristics as the GABAergic parvalbumin-positive neurons in the hippocampal formation[102] and in the cortex.[101] The high convergence and efficacy of CA3 inputs to parvalbumin-immunoreactive interneurons may at least in part be responsible for the high spontaneous firing rate that is observed in these neurons.[95,98,99,102,103]

Considering this high electrical activity of parvalbumin-immunoreactive neurons, the correlation of parvalbumin-immunoreactivity in brain regions with a highly enhanced cellular metabolism may be of relevance. In similar manner to the parvalbumin-immunoreactive auditory and visual nuclei[104,105] the parvalbumin-rich compartments in the visual cortex and in the dorsal lateral geniculate nucleus show enhanced activity of the enzyme cytochrome oxidase, whereas calbindin D-28K-immunoreactive neurons are located in areas with low enzymatic activities.[106-109] In the auditory and visual nuclei of the avian brain it could be demonstrated that these enhanced enzymatic activities are paired with an elevated electrical activity, as revealed by the ^{14}C-2-deoxyglucose method.[104,105] Assuming that bursting activity is a feature of most parvalbumin-immunoreactive neurons, this might at least in part account for the high metabolic (cytochrome oxidase) activity in these nuclei. Other possible causes for the enhanced enzymatic

activity are the highly energy-consuming intracellular events related to the various morphological alterations observed in these
areas after sensory or electrical stimulation and in the context of
learning-related neuronal plasticity (see below). For instance, protein synthesis, which is required for the formation and mainte-
nance of postsynaptic elements,[110] as well as microtubular transport in dendrites and axons, which is probably high during
formation and/or reduction of synapses, are energy consuming and
regulated by Ca^{2+}- and Mg^{2+}-ions.[111,112] In several studies, summarized in Table 5.1, it was shown that the expression of Ca^{2+}-bind-
ing proteins is regulated by electrical/metabolical activity levels.
After eye enucleation, cochlear extraction or even after social deprivation there is, depending on the protein and time course of the
investigation, an up- or down-regulation of protein content or
mRNA or a shift from the different metal-bound or metal-free
forms. So far, these studies provide at least indirect evidence for
the hypothesis that Ca^{2+}-binding proteins are in a still unknown
way involved in neuronal changes induced and regulated by different activation levels of the brain, which are altered in several
brain diseases, after injuries as well as during learning.

Related to this idea concerning the functional significance of
Ca^{2+}-binding proteins in certain neurons is the hypothesis that
these cells may be capable of producing Ca^{2+}-depolarizations.[113] The
subcellular localization of parvalbumin- and calbindin D-28K-
immunoreactivity mainly in somata and dendrites points to Ca^{2+}-
depolarizations chiefly in these compartments,[114] as has been
shown, for instance, in cerebellar Purkinje cells,[115-117] which contain parvalbumin as well as calbindin D-28K in their soma and
dendritic trees (see above). The recent ascertainment of an enhanced
density of voltage-dependent Ca^{2+} channels in parvalbumin- and,
in some cases, of calbindin D-28K-rich areas provides further support for this hypothesis. In some, but not all of the parvalbumin-
and calbindin D-28K-containing nuclei of the song birds' vocal
motor system as well as in a learning-relevant forebrain region of
chicks, there is good correlation with ligands for voltage-dependent Ca^{2+} channels such as the 1,2-dihydropyridine-receptor type
[125] J-Jodipine and the phenylalkylamine-receptor type ^{3}H-
desmethoxy-verapamil, which show enhanced binding in these regions.[118] A quite similar correlation of voltage-dependent Ca^{2+}-
channels with enhanced densities of parvalbumin- and/or calbindin

D-28K-positive structures can be found in the mammalian hippocampal formation. The highest densities of the 1,2-dihydropyridine receptor type of the voltage-dependent Ca^{2+}-channel have been found in the molecular layer of the dentate gyrus.[119-121] On the cellular level these ligands have been shown to be associated with the dendritic fields of the granule cells and with elements in the CA1 and CA3 regions of the hippocampus. Receptors of the phenylalkylamine type are enriched in the dentate gyrus, in the stratum oriens primarily of the CA3 region and in the subiculum.[122] While many CA1 neurons contain calbindin D-28K, the presence of parval-bumin and calbindin D-28K in CA3 neurons has been disputed.[86] Hence, the correlation of parvalbumin and calbindin D-28K immunoreactivity and voltage-dependent Ca^{2+} channels does not apply to all brain regions. Calcineurin has been proposed to interact with dihydropyridine-sensitive Ca^{2+} channels. By dephosphorylating the Ca^{2+} channel, or a closely associated protein, calcineurin may close the channel and thereby limit Ca^{2+} influx.[123]

There is also some electrophysiological evidence for the presence of Ca^{2+}currents in parvalbumin-positive neurons. Kawaguchi et al[102] found that parvalbumin-positive neurons in the CA1 region of the hippocampus belong to a class of fast-spiking neurons. Neuronal bursting activity seems to be mostly mediated by Ca^{2+}currents.[117,124] Bursting activity and the subsequent influx of Ca^{2+} ions seem to be instrumental in long-term potentiation, a commonly proposed model of the cellular learning mechanism.[124-128] Furthermore, it has been demonstrated that local application of excitatory amino acids like glutamate or NMDA induces long-lasting Ca^{2+}gradients in the apical dendrites of hippocampal CA1 neurons.[129]

Intracellularly, the depolarization-induced influx of Ca^{2+} ions into the neuron and subsequent interaction of the Ca^{2+} ions with Ca^{2+}-binding proteins might be part of a reaction cascade responsible, on the one hand, for fast changes in neuronal electrical properties and, on the other hand, for various long-term alterations of cell morphology that have been found to accompany learning and memory formation.[130-135]

Some evidence for an involvement in cellular alterations supposed to underly learning and memory formation has been gained for parvalbumin, calbindin D-28K, calmodulin, calcineurin and

S100 proteins.[136] For instance, in rat hippocampus, parvalbumin-immunoreactivity was dramatically decreased 24 hours after the animals had been trained on a brightness discrimination task;[137] this decrease was most pronounced in the CA1 region and was especially obvious in the parvalbumin-immunoreactive neuropil of this region. On the other hand, some authors argue that an impaired functioning of parvalbumin-immunoreactive neurons in the entorhinal cortex of aged rats may be responsible for the learning and memory deficits in these animals.[138] In aged rats, which show impaired performance in passive avoidance and water maze tasks, the number of parvalbumin-positive neurons was markedly decreased in the entorhinal, somatosensory and motor cortex as well as in the medial septum and vertical limb of the diagonal band of Broca, but not in the hippocampus. S100 proteins have been shown to be involved in some learning tasks in rats.[139,140] An increase in the S100 protein content in rat brains during training to acquire a new behavior has been reported.[141] Subdural injections of S100-antiserum did not change the EEG, but instead resulted in an impairment of maze running performance.[142] Intraventricular injections of an S100 antiserum inhibited further acquisition of new behavior in rats.[139]

In rat hippocampal slices, application of S100 antiserum into the stratum radiatum of area CA1 blocks the inducibility of long-term potentiation, an activity-induced increase of synaptic efficacy, which is assumed to be a physiological correlate for learning-related synaptic plasticity.[143] In addition, it has been shown that long-term potentiation cannot be induced in hippocampus slices when calmidazolium, an inhibitor of calmodulin, is applied.[144] Recently a role for calcineurin in the generation of long-term depression and in learning and memory formation has been proposed.[145] It was shown that calcineurin dephosphorylates and inactivates inhibitor 1, which in turn leads to an activation of postsynaptic serine/threonine protein phosphatase 1, which contributes to the generation of long-term depression.

On the cellular level there is some evidence for an interaction of certain Ca^{2+}-binding proteins with cytoskeletal elements, which may be relevant in the context of learning-related neuronal plasticity. Binding sites of calmodulin and actin on the brain calspectrin have been shown by rotary-shadowed electron microscopy.[146] Furthermore, calmodulin has been shown to affect cell morphology by changing the organization of cytoskeletal elements such as microfilaments, intermediate filaments and microtubules.[147-149] Calcineurin has also been shown to be associated with the cytoskeleton, and it may be involved in regulating the phosphorylation of tau protein.[150] Furthermore, calcineurin is enriched in growth cones where it may mediate Ca^{2+}-regulated enzymatic events and interactions between cytoskeletal systems during neurite elongation.[150] The visinin-like protein, VILIP, which is expressed in neurons and synapses of the chick brain has been shown to bind actin and thus may be another candidate for mediating synaptic plasticity.[151-153] For S100 proteins there is some evidence that they regulate microtubule assembly[154] and a possible role of S100β in glia-neuron interaction during the activity-dependent formation of cortical ocular dominance columns has been proposed.[155,156]

In addition to their interaction with the cytoskeleton, there is increasing evidence for an interaction of Ca^{2+}-binding proteins with components of the extracellular matrix.[157-165] The function of these extracellular matrix components is still unclear, they have been proposed to be involved in targeting glia-neuron interactions and stabilization of neurons and synapses.[164] Interestingly, many of the parvalbumin-immunoreactive neurons, but not the calretinin-immunoreactive neurons, are covered by "perineuronal nets" composed of extracellular matrix substances such as chondroitin sulfate, hyaluronan, tenascin and restrictin[161-163] (see Fig. 4.1D). It was speculated that this may protect the parvalbumin-positive neurons from excess exposure to excitatory amino acids, which are released during abnormal neuronal activity and ischemia, and thus enable them to survive under conditions that cause other neurons to degenerate and die.[159]

In the context of learning-related synaptic plasticity another aspect may be relevant, namely the ability of parvalbumin to influence the intracellular Mg^{2+}-concentration as well. Similarly as with Ca^{2+}, Mg^{2+} is essential for a variety of fundamental intracellular functions, including glycolysis, RNA- and DNA-synthesis, respiration and protein synthesis.[166-168] Protein synthesis in particular is very sensitive to alterations of intracellular Mg^{2+} concentrations.[168,169] In experimentally induced Mg^{2+} deficiency, an increase in cell permeability and consequently increased levels of intracellular Na⁺, Ca^{2+} and cAMP are observed, as well as an enhanced release of catecholamines.[170] Interestingly, if intracellular Mg^{2+} concentration in the brain is chronically elevated by a high Mg^{2+} diet, hippocampal frequency potentiation and reversal learning in aged and young rats are improved.[171]

In contrast to Ca^{2+}, which is assumed to play the specific role of the acute, transient regulation element, a complementary role was suggested for Mg^{2+}, namely as a more long-term regulatory element.[172] Thus, the regulation of intracellular Mg^{2+} concentrations might be another facet of the involvement of parvalbumin in neuronal plasticity.

In summary, the basically independent distribution and intracellular regulation patterns of these proteins as well as their widespread and even dominant occurrence in dendrites and synapses makes a special role for them in general mechanisms of chemical neurotransmission unlikely. Instead, their previously mentioned likely involvement in cellular plasticity and/or calcium potential regulation is given further support by their activity-dependent regulation found in many experimental studies (Table 5.1).

Finally, a role for these Ca^{2+}-binding proteins as protectors or as a "surviving factor" against neuronal degeneration and cell death is discussed in the context of neurodegenerative diseases. As mentioned above, cell death is accompanied by an enhanced intracellular Ca^{2+} concentration and some observations in human brain and in experimental animal models, which are in support of this hypothesis, will be described in chapter 6.

REFERENCES

1. Cohen P, Klee CB, eds. Calmodulin. Amsterdam, New York, Oxford: Elsevier, 1988.
2. Föhr UG, Weber BR, Müntener M et al. Human α and β parvalbumins. Structure and tissue-specific expression. Eur J Biochem 1993; 215:719-727.
3. Christakos S, Gabrielides C, Rhoten WB. Vitamin D-dependent calcium binding proteins: chemistry, distribution, functional considerations and molecular biology. Endocr Rev 1989; 10:3-26.
4. Heizmann CW, Braun K. Changes in Ca^{2+}-binding proteins in human neurodegenerative disorders. Trends Neurosci 1992; 15:259-264.
5. Lledo P-M, Somasundaram B, Morton AJ et al. Stable transfection of calbindin-D28k into the GH_3 cell line alters calcium currents and intracellular calcium homeostasis. Neuron 1992; 9:943-954.
6. Kosaka T, Kosaka K, Nakayama T et al. Axons and axon terminals

of cerebellar Purkinje cells and basket cells have higher levels of parvalbumin immunoreactivity than somata and dendrites: quantitative analysis by immunogold labeling. Exp Brain Res 1993; 93:483-491.

7. Dowd DR, MacDonald PN, Komm BS et al. Stable expression of the calbindin-D28K complementary DNA interferes with the apoptotic pathway in lymphocytes. Mol Endocrin 1992; 6: 1843-1848.

8. Heizmann CW. Parvalbumin, an intracellular calcium-binding protein: distribution, properties and possible roles in mammalian cells. Experientia 1984; 40:910-921.

9. Ushio H, Watabe S. Carp parvalbumin binds to and directly interacts with the sarcoplasmic reticulum for Ca²⁺ translocation. Biochem Biophys Res Commun 1994; 199:56-62.

10. Hou T, Johnson JD, Rall JA. Role of parvalbumin in relaxation of frog skeletal muscle. In: Sugi H, Pollack GH, eds. Mechanism of Myofilament Sliding in Muscle Contraction. New York: Plenum Press, 1993; 141-153.

11. Berridge MJ, Dupont G. Spatial and temporal signaling by calcium. Current Opinion in Cell Biol 1994; 6:267-274.

12. Tank DW, Sugimori M, Connor JA et al. Spatially resolved calcium dynamics of mammalian Purkinje cells in cerebellar slice. Science 1988; 242:773-777.

13. Chard PS, Bleakman D, Christakos S et al. Calcium buffering properties of calbindin D28k and parvalbumin in rat sensory neurones. J Physiol 1993; 472:341-357.

14. Mattson MP, Rychlik B, Chu C et al. Evidence for calcium-reducing and excitoprotective roles for the calcium-binding protein calbindin-D28k in cultured hippocampal neurons. Neuron 1991; 6:41-51.

15. Roberts WM. Spatial calcium buffer in saccular hair cells. Nature 1993; 363:74-76.

16. Feher JJ, Fullmer CS, Wasserman RH. Role of facilitated diffusion of calcium by calbindin in intestinal calcium absorption. Am J Physiol 1992; 262:C517-C526.

17. Haiech J, Derancourt J, Pechère J-F, et al. Magnesium and calcium binding to parvalbumins: Evidence for differences between parvalbumins and an explanation of their relaxing function. Biochem 1979; 18:2752-2758.

18. Gillis JM, Thomason D, Lefevre J et al. Parvalbumins and muscle relaxation: a computer simulation study. J Muscle Res Cell Mot 1982; 3:377-398.

19. Hou T-T, Johnson JD, Rall JA. Parvalbumin content and calcium and magnesium dissociation rates correlated with changes in relaxation rate of frog muscle fibers. J Physiol 1991; 441:285-304.

20. Renner M, Danielson MA, Falke JJ. Kinetic control of Ca(II) sig-

naling: Tuning the ion dissociation rates of EF-hand Ca(II) binding sites. Proc Natl Acad Sci USA 1993; 90:6493-6497.

21. Moore B. A soluble protein characteristic of the nervous system. Biochem Biophys Res Commun 1965; 19:739-744.

22. Hilt DC, Kligman D. The S100 protein family: a biochemical and functional overview. In: Heizmann CW, ed. Novel Calcium-Binding Proteins. Fundamentals and Clinical Implications. Berlin: Springer-Verlag, 1991: 65-103.

23. Zimmer DB. Examination of the calcium-modulated protein S100α and its target proteins in adult and developing skeletal muscle. Cell Motility and Cytoskeleton 1991; 20:325-337.

24. Donato R. Perspectives in S-100 protein biology. Cell Calcium 1991; 12:713-726.

25. Van Eldik LJ, Griffin WST. S100β expression in Alzheimer's disease: Relation to neuropathology in brain regions. Biochim Biophys Acta 1994; 1223:398-403.

26. Baudier J, Haglid K, Haiech J et al. Zinc binding to human brain calcium binding proteins, calmodulin and S100β protein. Biochem Biophys Res Commun 1983; 114:1138-1146.

27. Filipek A, Heizmann CW, Kuznicki J. Calcyclin is a calcium- and zinc-binding protein. FEBS Lett 1990; 264:263-266.

28. Duncan A, Higgins J, Dunn RJ et al. Refined sublocalization of the human gene encoding the β subunit of the S-100 protein (S-100β) and confirmation of a subtle t(9;21) translocation using in situ hybridization. Cytogenet Cell Genet 1989; 50:234-235.

29. Baudier J, Cole RD. Reinvestigation of the sulfhydryl reactivity in bovine brain S-100β protein and the microtubule-associated tau proteins. Ca^{2+} stimulates disulfide cross-linking between the S-100β subunit and the microtubule-associated tau protein. Biochem 1988; 27:2728-2736.

30. Becker LE, Mito T, Takashima S et al. Association of phenotypic abnormalities of Down syndrome with an imbalance of genes on chromosome 21. APMIS 1993; Suppl 40, 101:57-70.

31. Reeves RH, Yao J, Crowley MR et al. Astrocytosis and axonal proliferation in the hippocampus of S100β transgenic mice. Proc Natl Acad Sci USA 1994; 91:5359-5363.

32. Friend WC, Clapoff S, Landry C et al. Cell-specific expression of high levels of human S100β in transgenic mouse brain is dependent on gene dosage. J Neurosci 1992; 12:4337-4346.

33. Gerlai R, Friend W, Becker L et al. Female transgenic mice carrying multiple copies of the human gene for S100β are hyperactive. Behavioural Brain Res 1993; 55:51-59.

34. Kojima K, Berger T, Lassmann H et al. Experimental autoimmune panencephalitis and uveoretinitis transferred to the Lewis rat by T lymphocytes specific for the S100β molecule, a calcium binding protein of astroglia. J Exp Med 1994; 180:817-829.

35. Shashona VE, Hesse GW, Moore BW. Proteins of the extracellular fluid: evidence for release of S100 protein. J. Neurochem 1984; 42:1536-1541.

36. Zimmer DB, Van Eldik LJ. Levels and distribution of the calcium-modulated proteins S-100 and calmodulin in rat C 6 glioma cells. J. Neurochem 1988; 50:572-579.

37. Kligman D, Marshak D. Purification and characterization of a neurite extension factor from bovine brain. Proc Natl Acad Sci USA 1985; 82:7136-7139.

38. Winningham-Major F, Staecker JL, Barger SW et al. Neurite extension and neuronal survival activities of recombinant S-100β proteins that differ in the content and position of cysteine residues. J. Cell Biol 1989; 109:3063-3071.

39. Ueda S, Hou XA, Whitaker-Azmitia PM et al. Neuro-glial neurotrophic interaction in the S-100β retarded mutant mouse (Polydactyly Nagoya). II. Co-cultures study. Brain Res 1994; 633:284-288.

40. Mariggio MA, Fulle S, Calissano P et al. The brain protein S-100β induces apoptosis in PC12 cells. Neurosci 1994; 60:29-35.

41. Barraclough R, Savin J, Dube SK et al. Molecular cloning and sequence of the gene for p9Ka. A cultured myoepithelial cell protein with strong homology to S-100, a calcium-binding protein. J Mol Biol 1987; 198:13-20.

42. Ebralidze A, Tulchinsky E, Grigorian M et al. Isolation and characterization of a gene specifically expressed in different metastatic cells and whose deduced gene product has a high degree of homology to a Ca²⁺-binding protein family. Genes Devel 1989; 3:1086-1093.

43. Davies BR, Davies M, Gibbs F et al. Induction of the metastatic phenotype by transfection of a benign rat mammary epithelial cell line with the gene for p9Ka, a rat calcium-binding protein, but not with the oncogene EJ-ras-1. Oncogene 1993; 8:999-1008.

44. Jackson-Grusby LL, Siergiel J, Linzer DIH. A growth-related mRNA in cultured mouse cells encodes a placental calcium-binding protein. Nucleic Acids Res 1987; 15:667-669.

45. Watanabe Y, Usada N, Minami H. et al. Calvasculin, as a factor affecting the microfilament assemblies in rat fibroblasts transfected by src gene. FEBS Lett 1993; 324:51-55.

46. De Vouge MW, Mukherjee BB. Transformation of normal rat kidney cells by v-K-ras enhances expression of transin 2 and an S-100-related calcium-binding protein. Oncogene 1992; 7:109-119.

47. Lakshmi MS, Parker C, Sherbet GV. Metastasis-associated MTS1 and NM23 genes affect Tubulin polymerisation in B16 melanomas - a possible mechanism of their regulation of metastatic behavior of tumors. Anticancer Res 1993; 13:299-303.

48. Calabretta B, Battini R, Kaczmarek L et al. Molecular cloning of the cDNA for a growth-factor-inducible gene with strong homology to S-100, a calcium-binding protein. J Biol Chem. 1986; 261:12628-12632.

49. Wood L, Carter D, Mills M et al. Expression of calcyclin, a calcium-binding protein, in the keratogenous region of growing hair follicles. J Invest Derm 1991; 96:383-387.

50. Guo X, Chambers AF, Parfett CLJ et al. Identification of a serum-inducible messenger RNA (5B10) as the mouse homologue of calcyclin: tissue distribution and expression in metastatic, ras-transformed NIH 3T3 cells. Cell Growth Differ 1990; 1:333-338.

51. Weterman MAJ, Stoopen GM, Muijen GNPv et al. Expression of calcyclin in human melanoma cell lines correlates with metastatic behavior in nude mice. Cancer Res 1992; 52:1291-1296.

52. Lee SW, Tomasetto C, Swisshelm K et al. Down-regulation of a member of the S100 gene family in mammary carcinoma cells and re-expression by azadeoxycytidine treatment. Proc Natl. Acad Sci 1992; 89:2504-2508.

53. Engelkamp D, Schäfer BW, Mattei MG et al. Six S100 genes are clustered on human chromosome-1q21 - identification of two genes coding for the two previously unreported calcium-binding proteins - S100D and S100E. Proc Natl. Acad Sci 1993; 90:6547-6551.

54. Schäfer BW, Wicki R, Engelkamp D et al. Isolation of a YAC clone covering a cluster of nine S100 genes on human chromosome 1q21: rationale for a new nomenclature of the S100 calcium-binding protein family. Genomics 1995; 25:638-643.

55. Kriajevska MV, Neira Cardenas M, Grigorian MS et al. Non-muscle myosin heavy chain as a possible target for protein encoded by metastasis-related mts-1 gene. J Biol Chem 1994; 269:19679-19682.

56. Takenaga K, Nakamura Y, Endo H et al. Involvement of S100-related calcium-binding protein pEL98 (or mts1) in cell motility and tumor cell invasion. Jpn J Cancer Res 1994; 85:831-839.

57. Borg A, Zhang QX, Olsson H et al. Chromosome-1 alterations in breast cancer-allelic loss on 1p and 1q is related to lymphogenic metastases and poor prognosis. Genes Chromosome Cancer 1992; 5:311-320.

58. Pedrocchi M, Schäfer BW, Mueller H et al. Expression of Ca^{2+}-binding proteins of the S100 family in malignant human breast-cancer cell lines and biopsy samples. Int J Cancer 1994; 57:684-690.

59. Chen L-C, Dollbaum C, Smith HS. Loss of heterozygosity on chromosome 1q in human breast cancer. Proc Natl Acad Sci 1989; 86:7204-7207.

60. Kosaka T, Katsumaru H, Hama K et al. GABAergic neurons containing the calcium-binding protein parvalbumin in the rat hippocampus and dentate gyrus. Brain Res 1987; 419:119-130.

61. Celio MR. Parvalbumin in most gamma-aminobutyric acid-contain-

ing neurons of the rat cerebral cortex. Science 1986; 231:995-997.

62. Kosaka T, Heizmann CW, Tateishi K et al. An aspect of the organizational principle of the gamma-aminobutyric system in the cerebral cortex. Brain Res 1987; 409:403-408.

63. Katsumaru H, Kisaka T, Heizmann CW et al. Immunocytochemical study of GABAergic neurons containing the calcium-binding protein parvalbumin in the rat hippocampus. Exp Brain Res 1988a; 72:347-362.

64. Katsumaru H, Kosaka T, Heizmann CW et al. Gap-junctions on GABAergic neurons containing the calcium-binding protein parvalbumin in the rat hippocampus (CA1 regions). Exp Brain Res 1988b; 72:363-370.

65. Stichel CC, Singer W, Heizmann CW. Light and electron microscopy immunocytochemical localization of parvalbumin in the dorsal lateral geniculate nucleus of the cat: evidence for coexistence with GABA. J Comp Neurol 1988; 268:29-37.

66. Kosaka T, Kosaka K, Heizmann CW et al. An aspect of the organization of the GABAergic system in the rat main olfactory bulb: laminar distribution of immunohistochemically defined sub-population of GABAergic neurons. Brain Res 1987; 411:373-378.

67. DiFiglia M, Christakos S, Aronin N. Ultrastructural localization of immunoreactive calbindin-D28k in the rat and monkey basal ganglia, including subcellular distribution with colloidal gold labeling. J Comp Neurol 1989; 279:653-665.

68. Hendry SHC, Jones EG, Emson PC et al. Two classes of cortical GABA neurons defined by differential calcium binding protein immunoreactivities. Exp Brain Res 1989; 76:467-472.

69. Miettinen R, Gulyas AI, Baimbridge KG et al. Calretinin is present in non-pyramidal cells of the rat hippocampus - II. Coexistence with other calcium binding proteins and GABA. Neurosci 1992; 48:29-43.

70. Röhrenbeck J, Wässle H, Heizmann CW. Immunocytochemical labeling of horizontal cells in mammalian retina using antibodies against calcium-binding proteins. Neurosci Lett 1987; 77:255-260.

71. Nitsch R, Leranth C. Calretinin immunoreactivity in the monkey hippocampal formation - II. Intrinsic GABAergic and hypothalamic non-GABAergic systems: an experimental tracing and co-existence study. Neurosci 1993; 55:797-812.

72. Gerfen CR, Baimbridge KG, Thibault J. The neostriatal mosaic. III. Biochemical and developmental dissociation of patch-matrix mesostriatal systems. J Neurosci 1987; 7:3935-3944.

73. Weiss JH, Koh JY, Baimbridge KG et al. Cortical neurons containing somatostatin- or parvalbumin-like immunoreactivity are atypically vulnerable to excitotoxic injury in vitro. Neurol 1990; 40:1288-1292.

74. Waldvogel HJ, Faull RLM. Compartmentalization of parvalbumin

immunoreactivity in the human striatum. Brain Res 1993; 610:311-316.

75. Hartley D, Neve R, Bryan J et al. Expression of parvalbumin in cultured cortical neurons using a herpes simplex virus (HSV-1) vector system enhances NMDA-induced neurotoxicity. Soc Neurosci Abstr 1993; 19:1344.

76. Baimbrigde KG, Kuo J. Calbindin D-28K protects against glutamate induced neurotoxicity in rat CA1 pyramidal neurons cultures. Soc Neurosci Abstr 1988; 14:1264.

77. Mattson MP, Rychlik B, Chu C et al. Evidence for calcium-reducing and excito-protective roles for the calcium-binding protein calbindin D28K in cultured hippocampal neurons. Neuron 1991; 6:41-51.

78. Winsky L, Jacobowitz DM. Purification, identification and regional localization of a brain-specific calretinin-like calcium-binding protein (protein 10). In: Heizmann CW, ed. Novel Calcium Binding Proteins: Fundamentals and Clinical Implications. Berlin: Springer Verlag, 1991:277-300.

79. van der Zee EA, de Jong GI, Strosberg AD et al. Parvalbumin-positive neurons in rat dorsal hippocampus contain muscarinic acetylcholine receptors. Brain Res Bull 1991; 27:697-700.

80. Van der Zee EA, Luiten PGM. GABAergic neurons of the rat dorsal hippocampus express muscarinic acetylcholine receptors. Brain Res Bull 1993; 32:601-609.

81. Batini C, Palestini M, Thomasset M et al. Cytoplasmic calcium buffer, calbindin D-28K, is regulated by excitatory amino acids. NeuroReport 1993; 4:927-930.

82. Lowenstein DH, Miles MF, Hatam F et al. Up-regulation of calbindin D28KmRNA in the rat hippocampus following focal stimulation of the perforant path. Neuron 1991; 6:627-633.

83. Miller JJ, Baimbridge KG. Biochemical and immunohistochemical correlates of kindling induced epilepsy: role of calcium-binding protein. Brain Res 1983; 278:322-326.

84. Baimbridge KG, Miller JJ. Hippocampal calcium-binding protein during commissural kindling-induced epileptogenesis: progressive decline and effects of anticonvulsants. Brain Res 1984; 324:85-90.

85. Baimbridge KG, Mody I, Miller JJ. Reduction of rat hippocampal calcium-binding protein following comissural, amygdala, septal, perforant path, and olfactory bulb kindling. Epilepsia 1985; 26:460-465.

86. Sloviter RS. Calcium-binding protein (calbindin-D28K) and parvalbumin immunocytochemistry: localization in the rat hippocampus with specific reference to the selective vulnerability of hippocampal neuron to seizure activity. J Comp Neurol 1989; 280:183-196.

87. Sloviter RS, Sollas AL, Barbaro NM et al. Calcium-binding pro-

tein (calbindin-D28k) and parvalbumin immunocytochemistry in the normal and epileptic human hippocampus. J Comp Neurol 1991; 308:381-396.

88. Kamphuis W, Huisman E, Wadman WJ et al. Kindling induced changes in parvalbumin immunoreactivity in rat hippocampus and its relation to long-term decrease in GABA-immunoreactivity. Brain Res 1989; 479:23-34.

89. Heinemann U, Hamon B. Calcium and epileptogenesis. Exp Brain Res 1986; 65:1-10.

90. Dichter MA, Ayala GF. Cellular mechanisms of epilepsy: a status report. Science 1987; 237:157-164.

91. Pfyffer GE, Faivre-Baumann A, Tixier-Vidal A et al. Developmental and functional studies of parvalbumin and calbindin D28k in hypothalamic neurons grown in serum-free medium. J Neurochem 987; 49:442-451.

92. Sonnenberg JL, Frantz GD, Lee S et al. Calcium binding protein (calbindin D28K) and glutamate decarboxylase gene expression after kindling induced seizures. Molec Brain Res 1991; 9:179-190.

93. Llinas R, Sugimori M. Electrophysiological properties of in vitro Purkinje cell dendrites in mammalian cerebellar slices. J Physiol (Lond.) 1980; 305:197-213.

94. Wasterlain CG, Farber DB. Kindling alters the calcium/calmodulin-dependent phosphorylation of synaptic plasma membrane proteins in rat hippocampus. Proc Natl Acad Sci USA 1984; 81:1253-1257.

95. Kawaguchi Y, Hama K. Fast spiking nonpyramidal cells in the CA3 region, dentate gyrus and subiculum of rats. Brain Res 1987; 425:351-355.

96. Kawaguchi Y, Hama K. Two subtypes of nonpyramidal cells in rat hippocampal formation identified by intracellular recording and HRP injection. Brain Res 1987; 411:190-195.

97. Kawaguchi Y, Hama K. Physiological heterogeneity of nonpyramidal cells in rat hippocampal CA1 region. Exp Brain Res 1988; 72:494-502.

98. Lacaille JC, Schwartzkroin PA. Stratum lacunosum-moleculare interneurons of hippocampal CA1 region. I. Intracellular response characteristics, synaptic responses and morphology. J Neurosci 1988a; 8:1400-1424.

99. Lacaille JC, Schwartzkroin PA. Stratum lacunosum-moleculare interneurons of hippocampal CA1 region. II. Intrasomatic and intradendritic recordings of local circuit synaptic interactions. J Neurosci 1988b; 8:1411-1424.

100. Kawaguchi Y. Physiological, morphological, and histochemical characterization of three classes of interneurons in rat neostriatum. J Neurosci 1993; 13(11):4908-4923.

101. Kawaguchi Y, Kubota Y. Correlation of physiological subgroupings

of nonpyramidal cells with parvalbumin- and calbindin D28k-im-munoreactive neurons in layer V of rat frontal cortex. J Neuro-physiol 1993; 70(1):387-396.

102. Kawaguchi Y, Katsumaru H, Kosaka T et al. Fast-spiking cells in rat hippocampus (CA1-region) contain the calcium-binding pro-tein parvalbumin. Brain Res 1987; 416:369-374.

103. Schwartzkroin PA, Mathers LH. Physiological and morphological identification of a nonpyramidal hippocampal cell type. Brain Res 1978; 157:1-10.

104. Braun K, Scheich H, Schachner M et al. Distribution of parvalbumin, cytochrome oxidase activity and ^{14}C-2-deoxyglucose uptake in the brain of the zebra finch. I. Auditory and vocal mo-tor systems. Cell Tissue Res 1985; 240:101-115.

105. Braun K, Scheich H, Schachner M et al. Distribution of parvalbumin, cytochrome oxidase activity and ^{14}C-2-deoxyglucose uptake in the brain of the zebra finch. II. Visual system. Cell Tis-sue Res 1985; 240:117-127.

106. Celio MR, Schärer L, Morrison JH et al. Calbindin immunoreac-tivity alternates with cytochrome c-oxidase-rich zones in some lay-ers of the primate visual cortex. Nature (Lond.) 1986; 323:715-717.

107. Tigges M, Tigges J. Parvalbumin immunoreactivity in the lat-eral geniculate nucleus of rhesus monkeys raised under monocu-lar and binocular deprivation conditions. Vis Neurosci 1993; 10:1043-1053.

108. Blümcke I, Weruaga E, Kasas S et al. Discrete reduction pat-terns of parvalbumin and calbindin D-28k immunoreactivity in the dorsal lateral geniculate nucleus and the striate cortex of adult macaque monkeys after monocular enucleation. Vis Neurosci 1994; 11:1-11.

109. Spatz WB, Illing RB, Vogt Weisenhorn DM. Distribution of cyto-chrome oxidase and parvalbumin in primary visual cortex of the adult and neonate monkey, Callithrix jacchus. J Comp Neurol 1994; 339:519-534.

110. Burry RW. Protein synthesis requirement for the formation of syn-aptic elements. Brain Res 1985; 344:109-119.

111. Suarez-Isla BA, Pelto DJ, Thompson JM et al. Blockers of calcium permeability inhibit neurite extension and formation of neuromus-cular synapses in cell culture. Devel Brain Res 1984; 14:263-270.

112. Kater SB, Mattson MP, Cohan C et al. Calcium regulation of the neuronal growth cone. Trends Neurosci 1988; 11:315-321.

113. Braun K. Calcium-binding proteins in avian and mammalian cen-tral nervous system: localization, development and possible func-tions. Progr Histochem Cytochem 1990; 21/1:1-64.

114. Zuschratter W, Scheich H, Heizmann CW. Ultrastructural local-ization of the calcium-binding protein parvalbumin in neurons of the song system of the zebra finch. Cell Tissue Res 1985; 241:77-83.

115. Llinas R. Electrophysiological properties of dendrites in central neurons. In: Kreuzberg GW, ed. Advances in Neurology. New York: Raven Press, 1975:12:1-13.
116. Llinas R, Hess R. Tetrodotoxin-resistant dendritic spikes in avian Purkinje cells. Proc Natl Acad Sci USA 1976; 73:2520-2523.
117. Llinas R, Hess R. The role of calcium in neuronal function: In: Schmitt FO, Worden FG, eds. The Neurosciences: Fourth Study Program. Massachussetts-London: MIT Press Cambridge, 1979: 555-571.
118. Scheich H, Braun K. Synaptic selection and calcium regulation: Common mechanisms of auditory filial imprinting and vocal learning in birds?. In: Barth F, ed: Proceedings of the German Zoological Society. Stuttgart, New York: Gustav Fischer Verlag, 1988:77-95.
119. Murphy KMM, Gould RJ, Snyder SH. Autoradiographic visualization of (³H)nitrendipine binding sites in rat brain: localization to synaptic zones. Eur J Pharmacol 1982; 81:517-519.
120. Quirion R. Autoradiographic localization of a calcium channel antagonist, (³H)nitrendipine, binding sites in rat brain. Neurosci Lett 1983; 36:267-271.
121. Ferry DR, Goll A, Rombusch M et al. The molecular pharmacology and structural features of calcium channels. Brit J Clin Pharmacol 1985; 20:2335-2465.
122. Mourre C, Cervera P, Lazdunski M. Autoradiographic analysis in rat brain of the postnatal ontogeny of voltage-dependent Na⁺ channels, Ca⁺⁺-dependent K⁺-channels and slow Ca⁺⁺ channels identified as receptors for tetrodotoxin, apamin and (-)desmethoxy-verapamil. Brain Res 1987; 417:21-32.
123. Armstrong DL. Calcium channel regulation by calcineurin, a Ca²⁺-activated phosphatase in mammalian brain. Trends Neurosci 1989; 12:117-122.
124. Dunwiddie TV, Lynch G. The relationship between extracellular calcium concentrations and the induction of hippocampal long-term potentiation. Brain Res 1979; 169:103-110.
125. Turner RW, Baimbridge KG, Miller JJ. Calcium-induced long-term potentiation in the hippocampus. Neurosci 1982; 7:1411-1416.
126. Bliss TVP, Dolphin AC. Where is the locus of long-term potentiation? In: Lynch G, McGaugh JL, Weinberger NM, eds. Neurobiology of Learning and Memory. New York-London: The Guilford Press, 1984:451-458.
127. Kuhnt U, Mihaly A, Joo F. Increased binding of calcium in the hippocampal slice during long-term potentiation. Neurosci Lett 1985; 53:149-154.
128. Smith SJ. Progress on LTP at hippocampal synapses: a postsynaptic Ca²⁺-trigger for memory storage? Trends Neurosci 1987; 10:142-144.
129. Connor JA, Wadman WJ, Hockberger PE et al. Sustained dendritic gradients of Ca²⁺ induced by excitatory amino acids in CA1 hippocampal neurons. Science 1988; 240:649-653.

130. Van Harreveld A, Fifkova A. Swelling of dendritic spines in the fascia dentata after stimulation of the perforant fibers as a mechanism of post-tetanic potentiation. Exp Neurol 1975; 49:736-749.

131. Fifkova E, Van Harreveld A. Long-lasting morphological changes in dendritic spines of dentate granule cells following stimulation of the entorhinal area. J Neurocytol 1977; 6:211-230.

132. Wenzel J, Kammerer E, Kirsche W et al. Electron microscopic and morphometric studies on synaptic plasticity in the hippocampus of the rat following conditioning. J Hirnforsch 1980; 21:647-654.

133. Greenough WT. Structural correlates of information storage in the mammalian brain: a review and hypothesis. Trends Neurosci 1984; 7:229-233.

134. Greenough WT, Bailey CH. The anatomy of learning and memory: convergence of results across a diversity of tests. Trends Neurosci 1988; 11:142-147.

135. Lynch G, Baudry M. The biochemistry of memory: a new and specific hypothesis. Science 1984; 224:1057-1063.

136. Kasai H. Cytosolic Ca^{2+} gradients, Ca^{2+}-binding proteins and synaptic plasticity. Neurosci Res 1993; 16:1-7.

137. Handschack T, Zuschratter W, Staak S. Learning-related neuronal plasticity: changes in parvalbumin-immunoreactivity in the hippocampus of rats trained on a brightness discrimination task. Third European Symposium on Calcium Binding Proteins in Normal and Transformed Cells Zurich 1994; P3:42, Abstract.

138. Miettinen R, Sirviö J, Riekkinen P et al. Neocortical hippocampal and septal parvalbumin- and somatostatin-containing neurons in young and aged rats: correlation with passive avoidance and water maze performance. Neurosci 1993; 53(2):367-378.

139. Hyden H, Lange PW. The effect of antiserum to S100 protein on behavior and amount of S100 in brain cells. J Neurobiol 1981; 12:201-210.

140. Shashoua VE, Hesse GW, Moore BW. Proteins of the brain extracellular fluid: evidence for release of S-100 protein. J Neurochem 1984; 42:1536-1541.

141. Hyden H, Lange PW. S100 brain protein: correlation with behavior. Proc Natl Acad Sci USA 1970; 67:1959-1966.

142. Karpiak SE, Serokosz M, Rapport MM. Effects of antisera to S-100 protein and to synaptic membrane fraction on maze performance and EEG. Brain Res 1976; 102:313-321.

143. Lewis D, Teyler TJ. Anti-S100 serum blocks long-term potentiation in the hippocampal slice. Brain Res 1986; 383:159-164.

144. Reymann KG, Brödemann R, Kase H et al. Inhibitors of calmodulin and protein kinase C block different phases of hippocampal long-term potentiation. Brain Res 1988; 461:388-392.

145. Mulkey RM, Endo S, Shenolikar S et al. Involvement of a calcineurin/inhibitor-1 phosphatase cascade in hippocampal long-term depression. Nature 1994; 369:486-488.

146. Tsukita S, Tsukita S, Ishikawa H et al. Binding sites of calmodulin and actin on the brain spectrin, calspectin. J Cell Biol 1983; 97:574-578.
147. Kakiuchi S. Calmodulin-binding proteins in brain. Neurochem Internat 1983; 5:159-169.
148. Veigl ML, Vanaman TC, Sedwick WD. Calcium and calmodulin in cell growth and transformation. Biochem Biophys Acta (Amst.) 1984; 738:21-48.
149. Rasmussen CD, Means AR. Increased calmodulin affects cell morphology and mRNA levels of cytoskeletal protein genes. Cell Motility and the Cytoskeleton 1992; 21:45-57.
150. Ferreira A, Kincaid R, Kosik KS. Calcineurin is associated with the cytoskeleton of cultured neurons and has a role in the acquisition of polarity. Molecular Biology of the Cell 1993; 4:1225-1238.
151. Lenz SE, Braun K, Braunewell KH et al. VILIP - Ca²⁺ -dependent interaction with cell membrane and cytoskeleton. J Neurochem 1994; 63(1):72.
152. Braunewell KH, Lenz SE, Gundelfinger ED. VILIP-a 22kD neuronal EF-hand Ca²⁺-binding protein from chick brain: regulation of its interaction with intracellular target molecules. Soc Neurosci Abstr 1994; 20:1116.
153. Braunewell KH, Lenz SE, Gundelfinger ED. VILIP - a neuronal calcium binding protein: interaction with intracellular target molecules and possible involvement in neuronal differentiation. Biochemistry Hoppe-Seyler, Suppl. (in press).
154. Hesketh J, Baudier J. Evidence that S-100 proteins regulate microtubule assembly and stability in rat brain extracts. Int J Biochem 1986; 18:691-695.
155. Dyck RH, Van Eldik LJ, Cynader MS. Immunohistochemical localization of the S100β protein in postnatal cat visual cortex: spatial and temporal patterns of expression in cortical and subcortical glia. Devel Brain Res 1993; 72:181-192.
156. Müller ChM, Akhavan AC, Bette M. Possible role of S-100 in glia-neuronal signalling involved in activity-dependent plasticity in the developing mammalian cortex. J Chem Neuroanat 1993; 6:215-227.
157. Kosaka T, Heizmann CW. Selective staining of a population of parvalbumin-containing GABAergic neurons in the rat cerebral cortex by lectins with specific affinity for terminal N-acetylgalactosamine. Brain Res 1989; 483:158-163.
158. Drake CT, Mulligan KA, Wimpey TL et al. Characterization of Vicia villosa agglutinin-labeled GABAergic interneurons in the hippocampal formation and in acutely dissociated hippocampus. Brain Res 1991; 554:176-185.
159. Celio MR. Perineuronal nets of extracellular matrix around parvalbumin-containing neurons of the hippocampus. Hippocampus 1993; 3:55-60.

160. Brauer K, Härtig W, Bigl V et al. Distribution of parvalbumin-containing neurons and lectin-binding perineuronal nets in the rat basal forebrain. Brain Res 1993; 631:167-170.

161. Celio MR, Chiquet-Ehrismann R. "Perineuronal nets" around cortical interneurons expressing parvalbumin are rich in tenascin. Neurosci Lett 1993; 162:137-140.

162. Härtig W, Brauer K, Brückner G. Wisteria floribunda agglutinin-labeled nets surround parvalbumin-containing neurons. NeuroReport 1992; 3:869-872.

163. Härtig W, Brauer K, Bigl V et al. Chondroitin-sulfate proteoglycane-immunoreactivity of lectin-labeled perineuronal nets around parvalbumin-containing neurons. Brain Res 1994; 635:307-311.

164. Celio MR, Blümcke I. Perineuronal nets - a specialized form of extracellular matrix in the adult nervous system. Brain Res Rev 1994; 19:128-145.

165. Ren JQ, Heizmann CW, Kosaka T. Regional difference in the distribution of parvalbumin-containing neurons immunoreactive for monoclonal antibody HNK-1 in the mouse cerebral cortex. Neurosci Lett 1994; 166:221-225.

166. Günther T, Vormann J, Förster R. Functional compartmentation of intracellular magnesium. Magnesium-Bulletin 1984; 2:77-81.

167. Ferment O, Touitou Y. Magnesium: metabolism and hormonal regulation in different species. Comp Biochem Physiol 1985; 82A:753-758.

168. Günther T. Functional compartmentation of intracellular magnesium. Magnesium 1986; 5:53-59.

169. Terasaki M, Rubin H. Evidence that intracellular magnesium is present in cells at a regulatory concentration for protein synthesis. Proc Natl Acad Sci USA 1985; 82:7324-7326.

170. Günther T. Biochemistry and pathobiochemistry of magnesium. Artery 1981; 9:167-181.

171. Langfield PW, Morgan GA. Chronically elevating plasma Mg^{2+} improves hippocampal frequency potentiation and reversal learning in aged and young rats. Brain Res 1984; 322:167-171.

172. Grubbs RD, Maguire ME. Magnesium as a regulatory cation: criteria and evaluation. Magnesium 1987; 6:113-127.

CHAPTER 6

EF-Hand Ca²⁺-Binding Proteins in Neurodegenerative Disorders and their Use as Diagnostic Tools

NEURODEGENERATIVE DISORDERS

In human brain, parvalbumin, calbindin D-28K, calmodulin, calcineurin and S100 protein have been biochemically characterized (chapter 2) and immunocytochemically localized (chapter 4). Particular interest has been given to alterations of these proteins and the morphological changes occurring in cells containing Ca^{2+}-binding proteins as a consequence of neurological diseases (see Table 2.1).

6.1. EPILEPSY AND ISCHEMIA

HUMAN PATHOLOGY

The increase in intracellular calcium that occurs as a result of excitatory amino acid receptor activation has been suggested to be the initiating factor in seizure-associated degeneration and neuronal death.[1] Ca^{2+} overload is supposed to activate biochemical processes leading to enzymatic breakdown of proteins and lipids, to malfunctioning of mitochondria, energy failure and ultimately to cell death.[2] Since only certain subsets of neurons are susceptible to irreversible damage, the positive correlation between the content of a Ca^{2+}-binding protein and relative resistance to seizure-induced neuronal damage in certain hippocampal neuron populations is in support of the neuroprotective hypothesis of Ca^{2+}-binding proteins (chapter 5).[3,4,5] It is reasonable to assume that neurons containing certain intracellular Ca^{2+}-binding proteins and therefore having a greater capacity to buffer calcium, would be more resistant to degeneration. However, investigations about the vulnerability of such neurons in human brain as well as in experimental animal models revealed contradictory results concerning the postulated protective role of Ca^{2+}-binding proteins.

In human epileptic brain tissue there is a preferential survival of parvalbumin- and calbindin D-28K-immunoreactive neurons in the hippocampus.[4] In the dentate gyrus, most of the few surviving hilar neurons are calbindin D-28K-positive and only some calbindin D-28K-positive granule cells are lost. Other authors find a clear loss of parvalbumin- and calbindin D-28K-immunoreactive neurons.[6] One reason for these contradictory findings could be the different forms of epilepsies, which have been shown to display a remarkable variety of changes in density, distribution, staining intensity, size and morphological appearance of parvalbumin- and calbindin D-28K-immunoreactive neurons.[7] A more detailed study of parvalbumin-, calbindin D-28K- and calretinin-immunoreactive neurons in human epileptic tissue revealed that neuronal loss was generally associated with a decrease in calbindin D-28K-immunoreactive neurons.[8] These authors report that many of the surviving parvalbumin- and calbindin D-28K-immunoreactive neurons are covered with perineuronal nets, which are assumed to protect neurons from excitatory neurotransmitters. In contrast, none of the surviving calretinin-immunoreactive neurons were covered with perineuronal nets, but in some patients an increase of neuropil staining in the dentate gyrus and CA3 region as well as calretinin-immunostaining in many of the normally calretinin-negative granule cells was observed. Thus, the calretinin-positive neurons may be part of an activated and reorganizing system in epileptic hippocampus.

ANIMAL MODELS

One of the most intensively studied experimental models for epilepsy is the "kindling" stimulation. Kindling refers to a phenomenon in which repeated administration of an initially subconvulsive electrical stimulus results in progressive intensification of seizure activity, culminating in a generalized seizure. In this model, stimulation of the perforant path leads to a reduction only of parvalbumin-immunoreactive dentate basket cells,[4] but if kindling stimulation is applied, e.g. in the commissural fibers, the number of parvalbumin-positive neurons and the density of parvalbumin-positive neuropil increases.[9] While stimulation of the perforant path leads to a transient increase of calbindin D-28K-mRNA,[10] amygdaloid or commissural kindling leaves calbindin D-28K-mRNA-levels unchanged, whereas calbindin D-28K-immunoreactivity of dentate granule cells is decreased.[11,12,13] In a model where seizures are induced in rats by application of different numbers of electroconvulsive shocks, a transient decrease of calbindin D-28K-immunoreactivity, one to two days after the last shock is observed in the dentate granule cells including their dendrites and the mossy fibers.[14] This reduction of immunoreactivity is proportional to the number of shocks given to the animal and returns to normal levels after longer survival periods. No significant change is observed in the calbindin D-28K-immunoreactive pyramidal neurons or interneurons in the CA1-region of the hippocampus in these animals.

Reasons for these contradictory results may be, among others, different species (rats, mice, seizure- or shock-prone breeds), different technical approaches (biochemical assays, immunocytochemistry, in situ hybridization), the different stimulation paradigms, as well as the different time windows during which the authors analyze their effects (short-term (15 minutes) versus long-term (several hours, days, weeks). Furthermore, stimulation of different afferent pathways excite different neuronal subpopulations. The degree of their excitation may also be modulated by their intrinsic connectivity and different transmitter sensitivity and these factors together may determine their vulnerability.

In this context, the results from another experimental model for epilepsy, the "kainate model", are of interest. In this model repeated intraventricular or intrahippocampal injections of the glutamate agonist, kainate, are used to induce seizures, which are accompanied by similar pathological changes as seen in human epileptic tissue. After unilateral intracerebroventricular injection of kainate, an increasing loss of parvalbumin-immunoreactive neurons is observed in all areas of the ipsilateral hippocampus, and in the contralateral CA1-region after one to three weeks survival time.[15] Despite the massive cell loss in the dentate gyrus no reduction of parvalbumin-positive neurons is observed, which is in accordance with the findings in epileptic human tissue.[4] The ultrastructural analysis after three days survival time revealed many degenerating parvalbumin-immunoreactive neurons in the ipsi- and some in the contralateral CA1-region.[16] A few parvalbumin-positive neurons remain resistant to the kainate treatment, including their projecting terminals around the initial axon segments of the CA1 pyramidal neurons. In contrast, there is an almost complete loss of parvalbumin-positive terminals around the soma of the ipsilateral CA1 pyramidal neurons, the origin of which are the degenerated parvalbumin-positive neurons. Kainate injections directly into the CA3 area, which results in a massive degeneration of ipsi- and contralateral CA3 pyramidal neurons and an almost complete loss of contralateral CA1 pyramidal neurons, do not affect parvalbumin- and calbindin D-28K-immunoreactive non-pyramidal neurons.[17] In contrast to the neurons in the ipsi- and contralateral dentate gyrus, the spiny calretinin-immunoreactive neurons in the hilus and in CA3 are almost completely lost, which is in marked contrast to their relative resistance against kainate described in vitro.[18] Thus, like the somatostatin-immunoreactive neurons in the hilus the calretinin-positive neurons represent the most sensitive neurons in this experimental model. Calbindin D-28K-mRNA expression after kainate injection shows a transient increase in the hippocampus after 6 hours, which returns to baseline levels after 24 hours.[19] In an in vitro model of hippocampal slice cultures, 24h or 48h treatment with a dose of kainate, which induces persistent epileptiform activity, results in a complete loss of parvalbumin-immunoreactive interneurons after 3 days survival time.[20]

A selective sparing of the CA2 region has been found in an experimental model of epilepsy involving a combined ischemic and excitotoxic lesion to the rat hippocampus.[21] A certain degree of selective sparing of CA2 has also been indicated in experimental excitatory amino acid toxicity in vitro,[22] whereas in vivo all subfields of the dorsal hippocampus were equally affected by focal

QUESTIONNAIRE

Receive a FREE BOOK of your choice

Please help us out—Just answer the questions below, then select the book of your choice from the list on the back and return this card.

R.G. Landes Company publishes five book series: *Medical Intelligence Unit, Molecular Biology Intelligence Unit, Neuroscience Intelligence Unit, Tissue Engineering Intelligence Unit* and *Biotechnology Intelligence Unit.* We also publish comprehensive, shorter than book-length reports on well-circumscribed topics in molecular biology and medicine. The authors of our books and reports are acknowledged leaders in their fields and the topics are unique. Almost without exception, there are no other comprehensive publications on these topics.

Our goal is to publish material in important and rapidly changing areas of bioscience for sophisticated scientists. To achieve this goal, we have accelerated our publishing program to conform to the fast pace in which information grows in bioscience. Most of our books and reports are published within 90 to 120 days of receipt of the manuscript.

Please circle your response to the questions below.

1. We would like to sell our *books* to scientists and students at a deep discount. But we can only do this as part of a prepaid subscription program. The retail price range for our books is $59-$99. Would you pay $196 to select four *books* per year from any of our Intelligence Units–$49 per book–as part of a prepaid program?

    Yes   No

2. We would like to sell our *reports* to scientists and students at a deep discount. But we can only do this as part of a prepaid subscription program. The retail price range for our reports is $39-$59. Would you pay $145 to select five *reports* per year–$29 per report–as part of a prepaid program?

    Yes   No

3. Would you pay $39–the retail price range of our books is $59-$99–to receive any single book in our Intelligence Units if it is spiral bound, but in every other way identical to the more expensive hardcover version?

    Yes   No

To receive your free book, please fill out the shipping information below, select your free book choice from the list on the back of this survey and mail this card to:

R.G. Landes Company, 909 S. Pine Street, Georgetown, Texas 78626 U.S.A.

Your Name _____

Address _____

City_____ State/Province:_____

Country: _____ Postal Code:_____

My computer type is Macintosh_____ ; IBM-compatible _____ ; Other _____

Do you own ____ or plan to purchase ____ a CD-ROM drive?

AVAILABLE FREE TITLES

Please check three titles in order of preference.
Your request will be filled based on availability. Thank you.

☐ Water Channels
Alan Verkman,
University of California-San Francisco

☐ The Na,K-ATPase:
Structure-Function Relationship
J.-D. Horisberger, University of Lausanne

☐ Intrathymic Development of T Cells
J. Nikolic-Zugic,
Memorial Sloan-Kettering Cancer Center

☐ Cyclic GMP
Thomas Lincoln, University of Alabama

☐ Primordial VRM System and the Evolution
of Vertebrate Immunity
John Stewart, Institut Pasteur-Paris

☐ Thyroid Hormone Regulation
of Gene Expression
Graham R. Williams, University of Birmingham

☐ Mechanisms of Immunological Self Tolerance
Guido Kroemer, CNRS Génétique Moléculaire et
Biologie du Développement-Villejuif

☐ The Costimulatory Pathway
for T Cell Responses
Yang Liu, New York University

☐ Molecular Genetics of Drosophila Oogenesis
Paul F. Lasko, McGill University

☐ Mechanism of Steroid Hormone Regulation
of Gene Transcription
M.-J. Tsai & Bert W. O'Malley, Baylor University

☐ Liver Gene Expression
François Tronche & Moshe Yaniv,
Institut Pasteur-Paris

☐ RNA Polymerase III Transcription
R.J. White, University of Cambridge

☐ src Family of Tyrosine Kinases in Leukocytes
Tomas Mustelin, La Jolla Institute

☐ MHC Antigens and NK Cells
Rafael Solana & Jose Peña,
University of Córdoba

☐ Kinetic Modeling of Gene Expression
James L. Hargrove, University of Georgia

☐ PCR and the Analysis of the T Cell Receptor
Repertoire
Jorge Oksenberg, Michael Panzara & Lawrence
Steinman, Stanford University

☐ Myointimal Hyperplasia
Philip Dobrin, Loyola University

☐ Transgenic Mice as an In Vivo Model
of Self-Reactivity
David Ferrick & Lisa DiMolfetto-Landon,
University of California-Davis and Pamela Ohashi,
Ontario Cancer Institute

☐ Cytogenetics of Bone and Soft Tissue Tumors
Avery A. Sandberg, Genetrix & Julia A. Bridge ,
University of Nebraska

☐ The Th1-Th2 Paradigm and Transplantation
Robin Lowry, Emory University

☐ Phagocyte Production and Function Following
Thermal Injury
Verlyn Peterson & Daniel R. Ambruso,
University of Colorado

☐ Human T Lymphocyte Activation Deficiencies
José Regueiro, Carlos Rodríguez-Gallego
and Antonio Arnaiz-Villena,
Hospital 12 de Octubre-Madrid

☐ Monoclonal Antibody in Detection and
Treatment of Colon Cancer
Edward W. Martin, Jr., Ohio State University

☐ Enteric Physiology of the Transplanted Intestine
Michael Sarr & Nadey S. Hakim, Mayo Clinic

☐ Artificial Chordae in Mitral Valve Surgery
Claudio Zussa, S. Maria dei Battuti Hospital-Treviso

☐ Injury and Tumor Implantation
Satya Murthy & Edward Scanlon,
Northwestern University

☐ Support of the Acutely Failing Liver
A.A. Demetriou, Cedars-Sinai

☐ Reactive Metabolites of Oxygen and Nitrogen
in Biology and Medicine
Matthew Grisham, Louisiana State-Shreveport

☐ Biology of Lung Cancer
Adi Gazdar & Paul Carbone,
Southwestern Medical Center

☐ Quantitative Measurement
of Venous Incompetence
Paul S. van Bemmelen, Southern Illinois University
and John J. Bergan, Scripps Memorial Hospital

☐ Adhesion Molecules in Organ Transplants
Gustav Steinhoff, University of Kiel

☐ Purging in Bone Marrow Transplantation
Subhash C. Gulati,
Memorial Sloan-Kettering Cancer Center

☐ Trauma 2000: Strategies for the New Millennium
David J. Dries & Richard L. Gamelli,
Loyola University

injections of kainic acid and AMPA, while NMDA selectively affected CA1.[23] The resistance of CA2 to epileptic damage has been linked to the presence of high levels of Ca^{2+}-binding proteins in this field.[5,24] However, following chronic ethanol intoxication field CA2 shows the largest degree of neuronal cell loss, but it is unclear whether the mechanisms underlying cell death are similar.[25]

The striking similarity of the structural damage in conditions of epilepsy and ischemia has lead to the hypothesis that epileptic brain damage occurs as a result of cellular hypoxia.[2] Hypoxic cells have to cope with extremely high intracellular Ca^{2+} concentrations and those cells equipped with additional Ca^{2+} buffers may be less vulnerable. However, experimental data are yet somewhat inconsistent. For instance, while some authors report that parvalbumin-immunoreactive GABAergic interneurons in the gerbil hippocampus are resistant to ischemia-induced neuronal death,[26] others found a transient decrease of the number and staining intensity of parvalbumin-immunoreactive somata and fibers of CA1, CA3 and hilus after induction of ischemia, whereas in terminals parvalbumin-immunoreactivity remains unchanged.[27] Within days parvalbumin-immunoreactivity reappears. This points to a differential regulation of the protein within cellular compartments, which contain different concentrations of the protein. Other authors found a rapid and ephemeral increase of the intensity of parvalbumin-immunoreactivity shortly after the 20 minutes of ischemia in the CA1 area in gerbils, followed by a transitory decrease 1 hour later and again an increase after 6 hours in CA3 and the hilus and a delayed increase in CA1 one to two days later, even surpassing the number of immunoreactive neurons in CA1 in control animals.[28] When the delayed neuronal death in the CA1 area after 7 and 15 days is examined, the percentage of surviving parvalbumin-immunoreactive neurons is dramatically increased in ischemic animals compared to controls, suggesting that parvalbumin is associated with survival. Since there is some evidence from in vitro studies that the antiserum used in this study preferentially recognizes the Ca^{2+}-bound form of parvalbumin,[29] this initial increase of parvalbumin-immunoreactivity after ischemia is considered to reflect an increase of the Ca^{2+}-bound form of parvalbumin, which buffers the massive Ca^{2+} influx after ischemia, rather than an increase of parvalbumin synthesis. After long survival times (6-12 months) only a few parvalbumin-positive interneurons have survived, and their parvalbumin-immunoreactive neuropil is greatly reduced.[30] Calbindin D-28K-immunoreactivity in CA1 neurons is permanently lost, whereas calbindin D-28K-immunoreactivity in interneurons and the dentate granule cells remains almost unaffected.[27] In contrast, calbindin D-28K-mRNA expression after induction of ischemia shows a transient increase in the hippocampus after 6 hours returning to baseline levels after 24 hours.[31] In a study of developmental hypoxic-ischemic injury in young rats, the calbindin D-28K-immunoreactive neurons in the striatum seem to be less vulnerable to ischemic damage compared to surrounding neurons.[32] While the density of the intensively calbindin D-28K-immunoreactive neurons in the matrix compartment remains preserved, a prominent neuronal degeneration is observed in the striosome compartments, however with no changes in density of the faintly calbindin D-28K-immunoreactive neurons.

In contrast to the presumingly resistant parvalbumin- and calbindin D-28K-immunoreactive neurons, the hippocampal calretinin-immunoreactive neurons seem to be sensitive to ischemic damage. Particularly the spiny non-pyramidal neurons in the hilus and CA3 region, which are specifically associated with the mossy fiber system, are selectively lost already shortly after ischemia.[33] After one to three days survival time, the non-spiny and GABA-positive calretinin-immunoreactive neurons in CA1, CA3 and dentate gyrus have also disappeared. Assuming that the kainate- and AMPA-receptor type is predominant in the mossy fiber synapses on these calretinin-neurons, this finding is in accordance with similar observations after kainate injections.[17] In contrast, two weeks after ischemia the parvalbumin- and calbindin D-28K-immunoreactive non-pyramidal neurons in CA3 remain unchanged. Changes of another Ca^{2+}-binding protein, the Ca^{2+}-activated protease calpain and its endogenous inhibitor protein, calpastatin, are observed in response to the induction of hypoxia and hypoxic-ischemia.[34] While they are up-regulated in the hypoxic hemisphere in both the cytosolic and membrane-associated fraction, they are down-regulated in the ischemic-hypoxic hemisphere in the cytosolic fraction and increased in the membrane-associated fraction. Since lesions are not seen in the hypoxic hemisphere this up-regulation of calpain seems not to be necessarily coupled to development of injury.

On the one hand, this subtle intracellular regulation of levels of certain Ca^{2+}-binding proteins is suggestive for specialized mechanisms of intracellular Ca^{2+}-regulation and further supports the idea of a protective role of these proteins, which is also in good agreement with cell transfection experiments (Figs. 5.1 and 5.2). On the other hand, there are examples that demonstrate that the mere presence of a Ca^{2+}-binding protein is not sufficient to provide protection from degenerative processes. For instance, combined immunocytochemical and silver staining techniques to follow degeneration after transient ischemia in various brain regions revealed no consistent or systematic relationship between neuronal calbindin D-28K or parvalbumin content and vulnerability.[35,36] In addition, the calretinin-immunoreactive hilar neurons seem to be specifically vulnerable to ischemia and seizure related damage.[17] Furthermore, parvalbumin-immunoreactivity in interneurons of the hippocampal CA1 region is profoundly reduced after stroke, which eventually leads to a degeneration of these inhibitory neurons in spontaneously hypertensive stroke-prone rats.[37]

6.2. CHRONIC NEURODEGENERATIVE DISEASES

ALZHEIMER'S DISEASE

Impaired calcium homeostasis has been observed in Alzheimer's disease. Elevated calcium concentrations have been associated with neurons with Alzheimer-type neurofibrillary degeneration,[38] suggesting that altered calcium homeostasis may play a role in neuronal degeneration associated with Alzheimer's disease. Again, the results from studies on human pathologic material are inconsistent. For instance, some authors have found no changes in parvalbumin- and calretinin-immunoreactive neurons in prefrontal, frontal and inferior temporal cortex,[39-44] while other authors report a decrease in number and cell size of parvalbumin-immunoreactive neurons in temporal, parahippocampal, parietal and frontal cortex.[45,46] Parvalbumin-immunoreactive neurons are lost only in certain subfields such as CA3, subiculum and presubiculum, and in the prefrontal cortex only the calbindin D-28K interneurons in layers V-VI and pyramidal cells in the deep layer III are reduced, whereas the calbindin D-28K interneurons in layer II and upper layer III are unaffected.[40] More detailed analysis of the parvalbumin-immunoreactive neurons reveals that the degenerative processes mainly affect the neurites and dendrites of these neurons but not the parent neuron itself. For instance, in the temporal cortex and frontal lobe of Alzheimer patients there is no difference in the numbers or layer-specific distribution and density of parvalbumin-positive neurons.[43,47] However, in the temporal cortex a 35% loss of candlestick axons in layer II-III was observed[47] and in the frontal lobe abnormal degenerative parvalbumin-positive neurites forming aberrant sprouts, which are in contact with amyloid deposits or lying within senile plaques, were reported.[43] Calretinin-immunoreactive neurons in the entorhinal cortex of Alzheimer cases have reduced dendritic trees and dystrophic calretinin-immunoreactive fibers are observed in the dentate gyrus and in the hilus.[48] In the hippocampus, the prefrontal and the inferior temporal cortex the general pattern and morphology of calretinin-immunoreactive neurons appears well preserved in Alzheimer tissue.[44,48] Since the incorporation of neurites into amyloid deposits is an important step in the formation of senile plaques, the incorporation of neurites containing parvalbumin, calbindin D-28K and chromogranin A into neocortical plaques was investigated in detail.[49] There is a consistent and significant ranking, i.e. chromogranin A-immunoreactive neurites are preferentially incorporated into plaques compared to parvalbumin-immunoreactive neurites, and the sparse parvalbumin-positive neurites were favored over calbindin D-28K-immunoreactive neurites. Calretinin-immunoreactive neurons do not contain neurofibrillary tangles and no systematic association of their fibers with senile plaques can be found.[48]

For calbindin D-28K an overall decrease measured by radioimmunoassay and immunocytochemistry has been found in temporal, parietal and frontal cortex of Alzheimer's disease brains.[50-52] Calbindin D-28K-immunoreactive GABAergic interneurons in the neocortex including prefrontal cortex, which are largely confined to layer II with some additional, lightly calbindin D-28K-positive pyramidal neurons in layers III and V, seem to be relatively resistant to degeneration in Alzheimer's disease, only the lightly stained pyramidal neurons seem to undergo degenerative alterations.[39,40] Radioimmunoassay of calbindin D-28K in brains of normal individuals revealed an average cerebral cortical gray content of 312 μg/g soluble protein, with only slight variations in content over the cerebral hemisphere.[50] In brains of Alzheimer patients all neocortical areas revealed a significant decrease in calbindin D-28K to an average of 228 µg/g soluble protein. This decrease, however, is not specific to Alzheimer's disease, since patients with dementia of a histopathological type other than Alzheimer's disease (Huntington's chorea, Pick's disease, severe brain atrophy, dialysis dementia) show a similar reduction in calbindin D-28K to an average content of 244 µg/g soluble protein. Thus, depression of calbindin D-28K occurs in several encephalopathies.

Other authors report unchanged calbindin D-28K-mRNA and protein levels in cortical areas and in the cerebellum of Alzheimer's disease patients, but a decrease in the striatum, hippocampus (only in CA2 but not in the dentate gyrus) and nucleus raphe dorsalis.[53-55] A comparison between brain tissue from patients with Alzheimer's and with Huntington's disease reveals no changes of calbindin D-28K-mRNA and calmodulin-mRNA in temporal cortex and cerebellum, but a decrease of calbindin D-28K-mRNA in the hippocampal subfields, CA1 and CA2, and a decrease of calmodulin-mRNA in CA2 in Alzheimer brains compared to Huntington brains.[54]

In rat hippocampus, especially in the CA1 and CA3 region, the expression of calbindin D-28K and calretinin but not of calreticulin changes during "normal" aging, whereas no notable changes are observed in the cerebellum of these animals.[56] In contrast, others report no changes of calbindin D-28K-mRNA in aging hippocampus.[53,57] This discrepancy may be explained (at least partly) by the observation that only the ventral and central parts of the hippocampus showed significant reduction of calbindin D-28K[56] whereas dorsal parts show relatively good preservation of calbindin D-28K levels.

In the hippocampus of Alzheimer's disease brains the calbindin D-28K-immunoreactivity of stellate cells is more intense, whereas it is unchanged in granule cells. Enhanced terminal staining is observed in the outer molecular layer and senile plaques are evident. Non-pyramidal neurons show perikaryal swelling but their dendritic trees are largely unchanged. Thus, calbindin D-28K-immunoreactive neurons seem to be only weakly affected by degeneration in the hippocampal formation, possibly due to their enhanced Ca²⁺-buffering capacity. On the other hand, in the cerebral cortex a decrease in calbindin D-28K in several neurological forms of dementia has been biochemically detected.[50]

Alterations in calmodulin content associated with Alzheimer's disease have been detected, but again, they vary in the different brain regions.[50] The calmodulin content in the normal cerebral cortex is on the average 635 µg/g soluble protein, the calmodulin content of the occipital gray matter being about 10% less than for frontal gray matter. Subcortical white matter contains about one half the calmodulin of cerebral gray matter, i.e. 303 µg/g soluble protein and brainstem structures contain intermediate concentrations. In the cerebral cortex of Alzheimer's disease patients the average calmodulin content is significantly less with 446 µg/g soluble protein. Only the samples from the occipital lobes of Alzheimer-affected brains are not significantly different from normal brains. Brain damage associated with several other neurological diseases (such as Huntington's chorea, progressive supragranular palsy, dialysis encephalopathy, multi-infarct dementia, Pick's disease and severe unclassified cortical atrophy) is not associated with a statistically significant alteration in cortical calmodulin content compared with neurologically normal cerebral cortex. The calmodulin content of isolated nuclei and cell bodies prepared from Alzheimer-affected cerebral cortex was not significantly different from similar control preparations. This is an indication that the observed reduction in calmodulin content in Alzheimer's disease probably occurs in protoplasmic processes, including axons and dendrites distal to the perikaryon.

Subcortical white matter in Alzheimer brains exhibits a significant reduction in calmodulin content (244 µg/g soluble protein) compared with control brains (303 µg/g soluble protein). In contrast, samples from the putamen, caudate nucleus, thalamus and cerebellum are not different from the controls.

Calpain-immunoreactivity is present in cells undergoing tangle formation and these neurons are reduced in Alzheimer's disease brains.[41] S100 proteins are increased in reactive astrocytes in the temporal lobe of Alzheimer patients.[58]

Again it seems that neurons that contain a given Ca^{2+}-binding protein do not necessarily react in similar ways. Technical reasons for the contradictory findings, especially in human material, are the use of different antibodies and fixation parameters, as well as different postmortem intervals and the patient's age and clinical history. Furthermore, pathologically altered afferent and intrinsic connectivity and transmitter sensitivity may account for the observed regional variability.

CEREBELLAR DISEASES

During "normal" aging large decreases of calbindin D-28K-mRNA have been reported in the cerebellum of rats, mice and man,[53,57] others found no significant changes in rats and man.[59] Immunocytochemical investigations in aging rats revealed no changes of calbindin D-28K, calretinin, S100A6 and calcyclin in the cerebellum.[56] Again, the discrepancy of results may be due to regional variations of the observed changes.

In patients with cerebellar diseases involving damage of Purkinje cells (e.g. multiple system atrophy, subacute cerebellar degeneration in association with lung cancer, Wernicke-Korsakoff) increasing levels of calbindin D-28K can be measured in the cerebrospinal fluid (CSF) during illness.[60] Since Purkinje cells are only weakly calbindin D-28K-immunoreactive in these patients it may be that degenerating Purkinje cells release the protein into the CSF, and under this assumption the presence of calbindin D-28K levels in CSF may indicate the progression of cerebellar atrophy (see below). A quite similar increase of calbindin D-28K in CSF has been found in other neurodegenerative diseases such as Parkinson's and Huntington's diseases, where the affected neurons in the substantia nigra and striatum show decreased calbindin D-28K-mRNA levels.[53]

OTHER NEURODEGENERATIVE DISEASES

In Huntington's disease, which is a progressive neurodegenerative disease inherited through an autosomal dominant gene, parvalbumin-immunoreactive cortical neurons are reported to be resistant to degeneration.[61] In contrast, regional variations of changes in the density of parvalbumin-immunoreactive neurons have been reported in this disease by Ferrer et al.[62] While in the frontal cortex and in the dorsal putamen the density of parvalbumin-positive neurons is significantly decreased, their density in the temporal and primary visual cortices as well as in the globus pallidus appear normal. In the striatal matrix compartment there is a substantial loss of calbindin D-28K-immunoreactive neurons and their projection axons into the globus pallidus and substantia nigra pars reticulata show a marked reduction in Huntington's disease brains.[63] Other authors report no changes of calbindin D-28K-expression in the basal ganglia of Parkinson's disease patients, but observe a disorganization of the woolly fibers in the globus pallidus in patients with progressive supranuclear palsy.[64] A loss and dysmorphic alterations of medium sized spiny calbindin D-28K-immunoreactive striatal matrix neurons as well as proliferative and degenerative changes are described in Huntington patients.[65] Proliferative changes, which are predominantly found in moderate grades of the disease, include prominent recurving of distal dendrites as well as increased numbers and size of dendritic spines. Degenerative changes, predominant in severe grades of the disease consist of truncated dendritic arborizations, dendritic swellings and marked spine loss. These observations indicate that the presence of calbindin D-28K or parvalbumin alone does not protect cortical and neostriatal neurons from degeneration and dying.

Recently it has been shown that many of the parvalbumin-positive aspiny interneurons in the striatum contain the mitochondrial enzyme, manganese superoxide dismutase 2, which is involved in the scavenging of superoxide-free radicals in cells.[66] This high level of superoxide-free radical scavengers in parvalbumin-positive interneurons could explain why these neurons survive striatal degeneration observed following global ischemia or in Huntington's disease.

In Pick's disease no changes in parvalbumin-immunoreactive neurons are observed in frontal and temporal cortex by Arai et al,[67] but, although not mentioned by these authors, it appears from their figures as if there were a reduction of the density of parvalbumin-immunoreactive neuropil. In contrast, Satoh et al[46] report a decrease of parvalbumin-immunoreactive neurons in temporal, parahippocampal, parietal and cerebellar cortex, which is comparable to their observations in Alzheimer's disease and supranuclear palsy brains.

In brains of Parkinson patients a dramatic reduction of calbindin D-28K-mRNA and protein levels is found in the substantia nigra, hippocampus and in nucleus raphe dorsalis,[53] but the pigmented calbindin D-28K-immunoreactive neurons in the substantia nigra of these patients seem to be selectively spared from degeneration.[68] In an experimental model, where Parkinson-like syndromes are induced by the neurotoxin 1-methyl-4-phenyl-1,2,3,6-tetrahydropyridine (MPTP) in monkeys, it was shown that the calbindin D-28K-immunoreactive/tyrosine hydroxylase (TH)-containing neurons in the substantia nigra pars compacta and in the ventral tegmental area are much less severely affected than the TH-immunoreactive/calbindin D-28K-negative neurons in the same areas and that most spared neurons in these regions display both calbindin D-28K and TH.[69] In a similar MPTP model a massive loss of calbindin D-28K-immunoreactive Purkinje cells is observed and the few surviving Purkinje cells of the cerebellum are calbindin D-28K-negative.[70] It remains unclear whether these changes are caused by conformational changes of calbindin D-28K due to increased binding of Ca²⁺ ions and which then is no longer recognized by the antibody or whether there is an actual decrease of protein content due to advanced degeneration of the remaining Purkinje cells.

In the frontal cortex of patients suffering from a non-Alzheimer dementia associated with amyotropic lateral sclerosis (ALS) a dramatic decrease of calbindin D-28K-immunoreactive neurons is observed with the remaining calbindin D-28K-positive neurons expressing short dendritic arbors.[71] No change in the number of parvalbumin-immunoreactive neurons is found in this tissue. This differential reaction of the two neuron populations may be due to their anatomical location, with the calbindin D-28K-positive cells in the upper layers with generally extensive degeneration and the parvalbumin-positive neurons in the lower layers with less pronounced neuronal degeneration.

In the temporal lobe of patients suffering from Creutzfeldt-Jacob disease massive abnormalities of parvalbumin-immunoreactive neurons are observed that include reduced and short, often fragmented and varicose dendrites.[72] These massive signs of degeneration have not been observed in other cases of dementia including Alzheimer's disease.

6.3. OTHER BRAIN MALFUNCTIONS

OPIATES

Chronic treatment with morphine has been reported to increase calcium levels in synaptosomes of striatum and this may be mediated by persistent activation of the NMDA receptor. Since calbindin D-28K-immunoreactive neurons are selectively abundant in the matrix compartment of the striatum but almost absent in the μ-opiate-enriched striosomal compartment and since activation of glutamatergic transmission influences the intracellular levels of calbindin D-28K in hippocampal neurons, it may be reasoned that chronic morphine treatment might alter the expression of Ca^{2+}-binding proteins preferentially in the striosome compartment. Five-day treatment with morphine results in an increase of calbindin D-28K-immunoreactivity in striatal matrix neurons and induces intense calbindin D-28K-immunoreactivity in neurons within the striosome compartment. Co-administration of the NMDA-receptor antagonist MK-801 blocks this effect.[73] Thus, the finding that morphine treatment increases calbindin D-28K-immunoreactivity in the μ-receptor enriched striosome compartment could be consistent with increased activation of NMDA receptors.

One and four hours after single injections of morphine calbindin D-28K-mRNA levels in the cerebellum are decreased and stay unchanged in the remaining brain, co-administration of the opiate receptor antagonist naloxone reversed the effect of morphine.[74] This effect is reversed in rats that had previously been chronically treated with morphine in that there were no changes in the cerebellum but a more than two-fold increase of calbindin D-28K-mRNA levels in the remaining brain.

SCHIZOPHRENIA

The prefrontal cortical area of schizophrenic patients shows an increase of calbindin D-28K-immunoreactive local circuit neurons in layers I-III by 51% and in layer V by 88%.[75] In addition, the mean density of calretinin-immunoreactive neurons is increased by 36% in layers I and II of the schizophrenic subjects. Since the cortical thickness in these patients is not significantly different from control brains this increase of calbindin D-28K- and calretinin-immunoreactive neurons may not be attributed solely to a loss of neurons or neuropil in these brain regions.

TRAUMA, INJURIES

In rat parietal cortex the parvalbumin/GABAergic interneurons are particularly vulnerable to traumatic brain injury,[76] which has been shown to be accompanied by prolonged accumulations of calcium.[77] Calbindin D-28K-mRNA levels do not change following traumatic injury, although they increase after kainate induced seizures and ischemia (see above).[19]

AIDS

In brains of AIDS patients the number and size of astrocytes expressing S100β protein is greatly increased.[58]

6.4 CA²⁺-BINDING PROTEINS AS DIAGNOSTIC TOOLS

Alterations in calcium homeostasis play an important role in several pathological processes and Ca²⁺-binding proteins (listed in Table 2.3) are involved. These proteins are now used as diagnostic tools in neurodegenerative disorders,[78] myocardial hypertrophy or ischemia,[79-81] and monitoring of tumor progression.[82-84]

Some examples where Ca²⁺-binding proteins were evaluated for use in diagnostics are given below. Calbindin D-28K (enriched in cerebellar neurons) and S100β (located in glial cells) were measured by a sensitive immunoassay in the serum and cerebrospinal fluid after cardiac arrest to test their ability as indicators of brain damage.[81,85,86] The concentrations of both proteins increased dramatically after circulatory arrest and reperfusion, indicating their possible use as markers for the evaluation of brain damage after an ischemic period. Calbindin D-28K was also proposed as a marker for Purkinje cell damage since levels of this protein were markedly elevated in patients with cerebellar diseases.[85]

S100β neurotrophic activity was found to be elevated 10- to 20-fold in extracts of temporal lobe from autopsy samples of Alzheimer's disease patients.[87] Similarities between Alzheimer's disease and Down's syndrome neuropathology were noticed, e.g. location of the S100β gene on chromosome 21q22 (a region associated with the Down's syndrome phenotype), an increased number of reactive S100-positive astrocytes in both diseases and an elevated expression of the S100β gene.[87,88] Since S100β is secreted from glial cells, it will be possible to measure S100β levels with a quantitative assay not only in tissue extracts but also in CSF of these patients.

A recent hypothesis links aggregated β-amyloid peptide, generated by alternative processing of β-amyloid precursor protein, to destabilization of neuronal Ca²⁺ homeostasis, apparently by disrupting the Ca²⁺ regulatory system in the plasma membrane.[89] There is increasing evidence that Ca²⁺-binding proteins are altered in the brain of Alzheimer patients[78] and may provide a useful biochemical confirmation of neurophysiological and behavioral indicators.

Fortunately, increasing evidence suggests that non-neuronal somatic cells such as lymphocytes and skin fibroblasts provide a valuable tool for studying cellular and molecular aspects of Alzheimer's disease,[90-92] so that it may be possible to develop a test for altered levels of Ca^{2+}-binding proteins in cultured skin fibroblasts, which will reflect changes in brain cells. Several biochemical abnormalities were found in cultured fibroblasts from Alzheimer patients, including alterations in cytosolic free calcium and calcium uptake.

However, when Föhr et al[93] investigated the expression of several S100 proteins as well as calbindin D-28K, α-parvalbumin and calcineurin A by Northern, Western and ^{45}Ca blot analysis, they found that the altered levels of some Ca^{2+}-binding proteins in Alzheimer's brain do not persist in cultured fibroblasts from these patients. If this approach is to be useful, more Ca^{2+}-binding proteins will have to be tested and more sensitive immunoassays will have to be developed in order to find consistent diagnostic laboratory test systems for this disease.

The quantification of Ca^{2+}-binding proteins in tissue extracts and body fluids is mostly performed by immunoassays, e.g. troponin for the detection of myocardial infarction.[94-98] For these assays antibodies must be prepared and specific assays must first be developed, all of which is time consuming.

Recently, mass spectrometry has become a powerful method for identifying and quantifying proteins and peptides based on their molecular mass. A further development is the combination of high-performance liquid chromatography (HPLC) with electrospray ionization-tandem mass spectrometry to fractionate, determine the mass and sequence small amounts of proteins in complex cell extracts and body fluids.[99] Mass spectrometry has also been used to elucidate the complete amino acid sequences of α- and β-parvalbumins from several species[100-102] and those of calbindin D9K,[103] calbindin D-28K[104] and annexins.[105] Mass spectrometry has the additional advantage of also detecting post-translational modifications occurring in Ca^{2+}-binding proteins[106-108] that will influence their biological behavior. These modifications may be altered in the diseased state, which could be easily detected by this method.

Recently, this method has been applied for a quantitative analysis of several S100 proteins in human heart[109] (illustrated in Figs. 6.1 and 6.2).

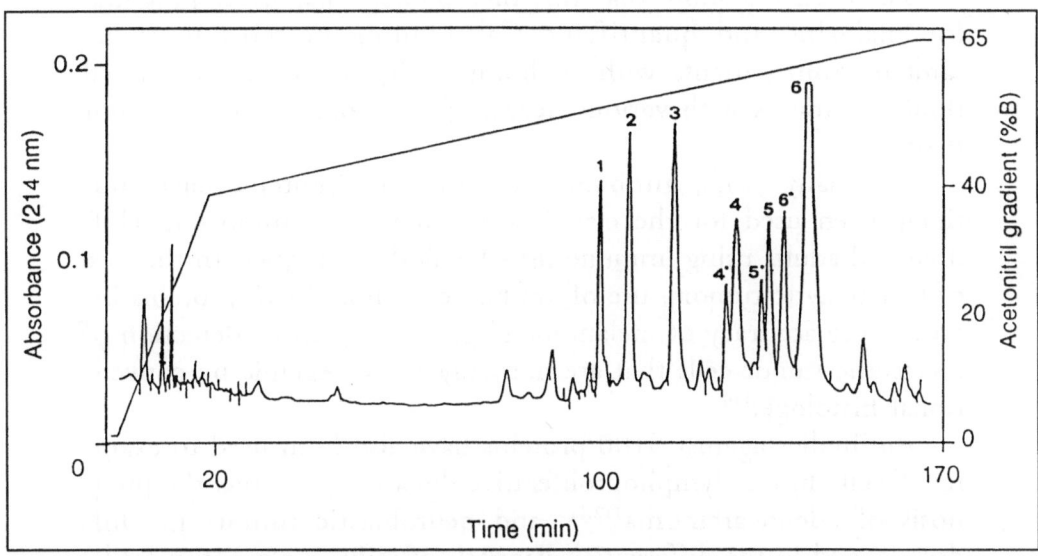

Fig. 6.1. Separation of S100 proteins from human heart by reverse phase high performance liquid chromatography (RP-HPLC). After extraction and ammonium sulfate fractionation, S100 proteins were first applied to a phenylsepharose column (not shown) and then eluted in the absence of Ca²⁺ (in the presence of EGTA). This eluate was then applied to the C8 RP-HPLC and 10 protein peaks were eluted by acetonitrile gradient and identified by electrospray ionization mass spectrometry (ESI-MS). Reprinted with permission from: Pedrocchi M, Hauer CR, Schäfer BW, et al. Biochem Biophys Res Commun 1993; 197:529-535; © Academic Press.

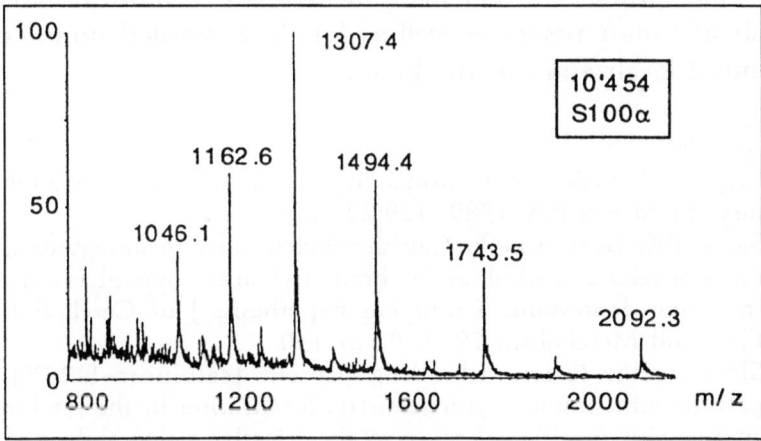

Fig. 6.2. ESI-MS spectrum of RP-HPLC peak no. 6. The HPLC peaks were directly injected into an ESI-MS to determine the molecular mass of, e.g. peak no. 4 of 10,454 kD, corresponding to S100A1 (S100α), (calculated mass from protein sequence, 10,458 kD); protein peak no. 5, S100A6 (calcyclin); peak no. 5, S100A4 (CAPL). Reprinted with permission from: Pedrocchi M, Hauer CR, Schäfer BW, et al. Biochem Biophys Res Commun 1993; 197:529-535; © Academic Press.

These results show that this methodology has great potential for analyzing and quantifying Ca^{2+}-binding proteins in tissue samples from patients with cardiomyopathy, in the cerebrospinal fluid of patients with various neurological disorders and in tumor tissues.

For many years, antibodies against S100 proteins have routinely been used for the classification of various tumors in children and adults using immunohistochemical techniques. In the case of melanocytic tumors, use of antibodies against S100 proteins increases the accuracy of melanoma staging and permits detection of metastatic tumor cells that are normally not detectable by conventional histology.[110]

Antibodies against S100 proteins have also been used to examine T cell chronic lymphoproliferative disease,[111] to assess the prognosis of adenocarcinoma[112,113] and neuroblastic tumors in children,[114] and in the differential diagnosis of salivary gland tumors.[115]

Commercially available antibodies produced against a mixture of bovine brain S100 proteins are only available for human tumor diagnostics.

Our laboratory has now produced specific antibodies against a number of recombinant human S100 proteins,[84,116,117] including the novel S100 proteins (listed in Table 2.2) for a more differential analysis of tumor tissues as well as for their detailed immunohistochemical localization in the brain.

REFERENCES

1. Meyer FB. Calcium, neuronal hyperexcitability and ischemic injury. Brain Res Rev 1989; 129:227-243.
2. Siesjö BK, Bengtsson F. Calcium-fluxes, calcium antagonists, and calcium-related pathology in brain ischemia, hypoglycemia, and spreading depression: a unifying hypothesis. J of Cerebral Blood Flow and Metabolism 1989; 9:127-140.
3. Sloviter RS. Calcium-binding protein (calbindin-D28K) and parvalbumin immunocytochemistry: localization in the rat hippocampus with specific reference to the selective vulnerability of hippocampal neurons to seizure activity. J Comp Neurol 1989; 280:183-196.
4. Sloviter RS, Sollas AL, Barbaro NM et al. Calcium-binding protein (calbindin-D28k) and parvalbumin immunocytochemistry in the normal and epileptic human hippocampus. J Comp Neurol 1991; 308:381-396.

5. Leranth C, Ribak CE. Calcium binding proteins are concentrated in the CA2 field of the monkey hippocampus: A possible key to this region's resistance to epileptic damage. Exp Brain Res 1991; 85:129-136.
6. Vonau M, Törk I. Soc Neurosci Abstr 1991; 17:1260.
7. Ferrer I, Oliver B, Russi A et al. Parvalbumin and calbindin D28K immunocytochemistry in human neocortical epileptic foci. J Neurol Sci 1994; 123:18-25.
8. Eickhoff C, Blümcke I, Celio MR, et al. Distribution of calcium-binding proteins parvalbumin, calbindin D-28K and calretinin, and perineuronal nets in the human epileptic hippocampus. Soc Neurosci Abstr 1994, 20:592.7.
9. Kamphuis W, Huisman E, Wadman WJ et al. Kindling induced changes in parvalbumin immunoreactivity in rat hippocampus and its relation to long-term decrease in GABA-immunoreactivity. Brain Res 1989; 479:23-34.
10. Lowenstein DH, Miles MF, Hatam F et al. Up-regulation of calbindin D28KmRNA in the rat hippocampus following focal stimulation of the perforant path. Neuron 1991; 6:627-633.
11. Baimbridge KG, Miller JJ. Hippocampal calcium-binding protein during commissural kindling-induced epileptogenesis: progressive decline and effects of anticonvulsants. Brain Res 1984; 324:85-90.
12. Baimbridge KG, Mody I, Miller JJ. Reduction of rat hippocampal calcium-binding protein following commissural, amygdala, septal, perforant path, and olfactory bulb kindling. Epilepsia 1985; 26:460-465.
13. Sonnenberg JL, Frantz GD, Lee S et al. Calcium binding protein (calbindin D28K) and glutamate decarboxylase gene expression after kindling induced seizures. Molec Brain Res 1991; 9:179-190.
14. Tonder N, Kragh J, Bolwig T et al. Transient decrease in calbindin immunoreactivity of rat fascia dentata granule cells after repeated electroconvulsive shocks. Hippocampus 1994; 4:79-84.
15. Best N, Mitchell J, Baimbridge KG et al. Changes in parvalbumin-immunoreactive neurons in the rat hippocampus following a kainic acid lesion. Neurosci Lett 1993; 155:1-6.
16. Best N, Mitchell J, Wheal HV. Ultrastructure of parvalbumin-immunoreactive neurons in the CA1 area of the rat hippocampus following a kainic acid injection. Acta Neuropathol 1994; 87:187-195.
17. Magloczky Z, Freund TF. Selective neuronal death in the contralateral hippocampus following unilateral kainate injections into the CA3 subfield. Neurosci 1993; 56:317-336.
18. Winsky L, Jacobowitz DM. Purification, identification and regional localization of a brain-specific calretinin-like calcium-binding protein (protein 10). In: Heizmann CW, ed. Novel Calcium Binding Proteins: Fundamentals and Clinical Implications. Berlin: Springer Verlag, 1991:277-300.

19. Lowenstein DH, Gwinn RP, Seren S et al. Increased expression of mRNA encoding calbindin D-28K, the glucose-regulated proteins, or the 72 kDA heat-shock protein in three models of acute CNS injury. Molec Brain Res 1994; 22:299-308.

20. Benedikz E, Casaccia-Bonnefil P, Stelzer A et al. Hyperexcitability and cell loss in kainate-treated hippocampal slice cultures. NeuroReport 1993; 5:90-92.

21. Franck JE, Roberts DL. Combined kainate and ischemia produces "mesial temporal sclerosis". Neurosci Lett 1990; 118:158-163.

22. Mattson MP, Guthrie PB, Kater SB. Instrinsic factors in the selective vulnerability of hippocampal pyramidal neurons. Prog Clin Biol Res 1989; 317:333-351.

23. Moncada C, Arvin B, Le Peillet E et al. Non-NMDA antagonists protect against kainate more than AMPA toxicity in the rat hippocampus. Neurosci Lett 1991; 133:287-290.

24. Munoz DG. The distribution of chromogranin A-like immunoreactivity in the human hippocampus coincides with the pattern of resistance to epilepsy-induced neuronal damage. Ann Neurol 1990; 27:266-275.

25. Bengoechea O, Gonzalo LM. Effects of alcoholization on the rat hippocampus. Neurosci Lett 1991; 123:112-114.

26. Nitsch C, Scotti A, Sommacal et al. GABAergic hippocampal neurons resistant to ischemia-induced neuronal death contain the Ca-binding protein parvalbumin. Neurosci Lett 1989; 105:263-268.

27. Johansen FF, Tonder N, Zimmer J et al. Short-term changes of parvalbumin and calbindin immunoreactivity in the rat hippocampus following cerebral ischemia. Neurosci Lett 1990; 120:171-174.

28. Tortosa A, Ferrer I. Parvalbumin immunoreactivity in the hippocampus of the gerbil after transient forebrain ischemia: a qualitative and quantitative sequential study. Neuroscience 1993; 1:33-43.

29. Pfyffer GE, Faivre-Baumann A, Tixier-Vidal A et al. Developmental and functional studies of parvalbumin and calbindin D28k in hypothalamic neurons grown in serum-free medium. J Neurochem 1987; 49: 442-451.

30. Mudrick LA, Baimbridge KG. Long-term structural changes in the rat hippocampal formation following cerebral ischemia. Brain Res 1989; 493:179-184.

31. Lowenstein DH, Gwinn RP, Seren S et al. Increased expression of mRNA encoding calbindin D-28K, the glucose-regulated proteins, or the 72 kDA heat-shock protein in three models of acute CNS injury. Molec Brain Res 1994; 22:299-308.

32. Burke RE, Baimbridge KG. Relative loss of the striatal striosome compartment, defined by calbindin-D28k immunostaining, following developmental hypoxic-ischemic injury. Neurosci 1993; 56:305-315.

33. Freund TF, Magloczky Z. Early degeneration of calretinin-containing neurons in the rat hippocampus after ischemia. Neurosci 1993; 56:581-596.
34. Ostwald K, Hagberg H, Andine P et al. Up-regulation of calpain activity in neonatal rat brain after hypoxic-ischemia. Brain Res 1993; 630:289-294.
35. Freund TF, Buszaki G, Leon A et al. Relationship of neuronal vulnerability and calcium binding protein immunoreactivity in ischemia. Exp Brain Res 1990; 83:55-66.
36. Freund TF, Ylinen A, Miettinen R et al. Pattern of neuronal death in the rat hippocampus after status epilepticus. Relationship to calcium binding protein content and ischemic vulnerability. Brain Res Bull 1991; 28:27-38.
37. De Jong GI, van der Zee EA, Bohus B et al. Reversed alterations of hippocampal parvalbumin and protein kinase C-immunoreactivity after stroke in spontaneously hypertensive stroke prone rats. Stroke 1993; 24:2082-2086.
38. Perl D, Brody AR. Alzheimer's disease: X-ray spectrometric evidence for aluminium accumulation in neurofibrillary tangle-bearing neurons. Science 1980; 208:297-215.
39. Morrison JH, Cox K, Hof PR et al. Neocortical parvalbumin-containing neurons are resistant to degeneration in Alzheimer's disease. Soc Neurosci Abstr 1988; 14:1085.
40. Hof PR, Morrison JH. Neocortical neuronal subpopulations labeled by a monoclonal antibody to calbindin exhibit differential vulnerability in Alzheimer's disease. Exp Neurol 1991; 111:293-301.
41. Iwamoto N, Emson P. Demonstration of neurofibrillary tangles in parvalbumin-immunoreactive interneurons in the cerebral cortex of Alzheimer-type dementia brain. Neurosci Lett 1991; 128:81-84.
42. Ferrer I, Soriano E, Tunon T et al. Parvalbumin-immunoreactive neurons in normal human temporal neocortex and in patients with Alzheimer's disease. J Neurol Sci 1991; 106:135-141.
43. Ferrer I, Zujar MJ, Rivera R, Soria M et al. Parvalbumin-immunoreactive dystrophic neurites and aberrant sprouts in the cerebral cortex of patients with Alzheimer's disease. Neurosci Lett 1993; 158:163-166.
44. Hof PR, Nimchinsky EA, Celio MR et al. Calretinin-immunoreactive neocortical interneurons are unaffected in Alzheimer's disease. Neurosci Lett. 1993; 152:145-149.
45. Arai H, Emson PC, Mountjoy CQ et al. Loss of parvalbumin-immunoreactive neurons from cortex in Alzheimer-type dementia. Brain Res. 1987; 418:164-169.
46. Satoh J, Tabira T, Sano M et al. Parvalbumin-immunoreactive neurons in the human central nervous system are decreased in Alzheimer's disease. Acta Neuropathol 1991; 81:388-395.

47. Fonseca M, Soriano E, Ferrer I et al. Chandelier cell axons identified by parvalbumin-immunoreactivity in the normal human temporal cortex and in Alzheimer's disease. Neuroscience 1993; 55:1107-1116.

48. Brion JP, Resibois A. A subset of calretinin-positive neurons are abnormal in Alzheimer's disease. Acta Neuropathol 1994; 88:33-43.

49. Adams LA, Munoz DG. Differential incorporation of processes derived from different classes of neurons into senile plaques in Alzheimer's disease. Acta Neuropathol 1993; 86:365-370.

50. Crapper McLachlan DR, Wong L, Bergeron C et al. Calmodulin and calbindin 28K in Alzheimer disease. Alzheimer Dis Assoc Disorders 1987; 1:179.

51. Ichimiya Y, Emson PC, Mountjoy CQ et al. Loss of calbindin-D28K immunoreactive neurons from the cortex in Alzheimer-type dementia. Brain Res 1988; 475:156-159.

52. Nishiyama E, Ohwada J, Iwamoto N et al. Selective loss of calbindin D-28K-immunoreactive neurons in the cortical layer II in brains of Alzheimer's disease: a morphometric study. Neurosci Lett 1993; 163:223-226.

53. Iacopino AM, Christakos S. Specific reduction of calcium-binding proteins (28-kilodalton calbindin-D) gene expression in aging mouse cerebellum. Proc Natl Acad Sci USA 1990; 87:4078-4082.

54. Sutherland MK, Wong L, Somerville MJ et al. Reduction of calbindin-28k mRNA levels in Alzheimer as compared to Huntington hippocampus. Molec Brain Res 1993; 18:32-42.

55. Maguire-Zeiss KA, Li ZW, Shimoda LMN et al. Calbindin-D28K mRNA in hippocampus, superior temporal gyrus and cerebellum: comparison between control and Alzheimer's disease subjects. Brain Res 1995; (in press).

56. Villa A, Podini P, Panzeri C et al. Cytosolic Ca^{2+}-binding proteins during rat brain aging: loss of calbindin and calretinin in the hippocampus, with no changes in the cerebellum. Eur J Neurosci 1994; 6:1491-1499.

57. Iacopino AM, Rhoten WB, Christakos S. Calcium binding proteins (calbindin-D28K) gene expression in the developing and aging mouse cerebellum. Molec Brain Res 1990; 8:283-290.

58. Griffin WST, Stanley LC, Ling C et al. Brain interleukin 1 and S100 immunoreactivity are elevated in Down syndrome and Alzheimer's disease. Proc Natl Acad Sci USA 1989; 86:7611-7615.

59. Kurobe N, Inaguma Y, Shinohara H et al. Developmental and age-dependent changes of 28-kDa calbindin D-28K in the central nervous tissue determined with a sensitive immunoassay method. J Neurochem 1992; 58:128-134.

60. Kiyosawa K, Mokuno K, Murakami N et al. Cerebrospinal fluid 28-k Da calbindin-D as a possible marker for Purkinje cell damage. J Neurol Sc 1993; 118:29-33.

61. Cudkowicz M, Kowall NW. Parvalbumin immunoreactive neurons are resistant to degeneration on Huntington's Disease cerebral cortex. J Neuropathol Exp Neurol 1990; 49:345.

62. Ferrer I, Kulisevsky J, Gonzales G et al. Parvalbumin-immunoreactive neurons in the cerebral cortex and striatum in Huntington's disease. Neurodegeneration 1994; 3:169-173.

63. Seto-Ohshima A, Emson PC, Lawson E et al. Loss of matrix calcium-binding protein-containing neurons in Huntington's disease. The Lancet 1988; 1:1252-1254.

64. Ito H, Goto S, Sakamoto S et al. Calbindin D-28K in the basal ganglia of patients with Parkinsonism. Ann Neurol 1991; 32:543-550.

65. Ferrante RJ, Kowall NW, Richardson EP Jr. Proliferative and degenerative changes in striatal spiny neurons in Huntington's disease: A combined study using the section-Golgi method and calbindin D28k immunocytochemistry. J Neurosci 1991; 11:3877-3887.

66. Reiner A, Fugueredo-Cardenas G, Medina L. Preferential expression of superoxide dismutase by cholinergic and parvalbumin neurons in the monkey striatum. Soc Neurosci Abstr 1994, 20:671.11.

67. Arai H, Noguchi I, Makino Y et al. Parvalbumin-immunoreactive neurons in the cortex in Pick's disease. J Neurol 1991; 238:200-202.

68. Yamada T, McGeer PL, Baimbridge KG et al. Relative sparing in Parkinson's disease of substantia nigra dopamine neurons containing calbindin D28K. Brain Res 1990; 526:303-307.

69. Lavoie B, Parent A. Dopaminergic neurons expressing calbindin in normal and Parkinsonian monkeys. NeuroReport 1991; 2:601-604.

70. Vignola C, Necchi D, Scherini E et al. MPTP-induced changes in the monkey cerebellum. Immunohistochemistry of calcium binding and cytoskeletal proteins. Neurodegeneration 1994; 3:25-31.

71. Ferrer I, Tunon T, Serrano MT et al. Calbindin D-28K and parvalbumin immunoreactivity in the frontal cortex in patients with frontal lobe dementia of non-Alzheimer type associated with amyotropic lateral sclerosis. J Neurol Neurosurg Psychiatry 1993; 56:257-261.

72. Ferrer I, Casas R, Rivera R. Parvalbumin-immunoreactive cortical neurons in Creutzfeldt-Jacob disease. Ann Neurol. 1993; 34:864-866.

73. Garcia MM, Harlan RE. Chronic morphine increases calbindin D28k in rat striatum: possible NMDA receptor involvement. NeuroReport 1993; 5:65-68.

74. Tirumalai PS, Howells RD. Regulation of calbindin D-28K gene expression in response to acute and chronic morphine administration. Molec Brain Res 1994; 23:144-150.

75. Daviss SR, Lewis DA. Calbindin- and calretinin-immunoreactive local circuit neurons are increased in density in the prefrontal cortex of schizophrenic subjects. Soc Neurosci Abstr 1993; 19:84.9.

76. Querido K, Lee SM, Hoyda DA et al. Fluid-percussion brain injury selectively destroys parvalbumin containing cells in rat parietal cortex. Soc Neurosci Abstr 1993; 19:1878.

77. Fineman I, Hovda DA, Smith M et al. Concussive brain injury is associated with a prolonged accumulation of calcium: a ^{45}Ca autoradiographic study. Brain Res 1993; 624:94-102.

78. Heizmann CW, Braun K. Changes in Ca^{2+}-binding proteins in human neurodegenerative disorders. Trends Neurosci 1992; 15:259-264.

79. Morgan JP. Abnormal intracellular modulation of calcium as a major cause of cardiac contractile dysfunction. New England J Med 1991; 325:625-632.

80. Quaife RA, Kohmoto O, Barry WH. Mechanisms of reoxygenation injury in cultured ventricular myocytes. Circulation 1991; 83:566-577.

81. Usui A, Kato K, Sasa H et al. S-100α_o protein in serum during acute myocardial infarction. Clin Chem 1990; 36:639-641.

82. Weterman MA, Stoopen GM, Muijen GNP v et al. Expression of calcyclin in human melanoma cell lines correlates with metastatic behaviour in nude mice. Cancer Res 1992; 52:1291-1296.

83. Davies BR, Davies M, Gibbs F et al. Induction of the metastatic phenotype by transfection of a benign rat mammary epithelial cell line with the gene for p9Ka, a rat calcium-binding protein, but not with the oncogene EJ-ras-1. Oncogene 1993; 8:999-1008.

84. Pedrocchi M, Schäfer BW, Mueller H et al. Expression of Ca^{2+}-binding proteins of the S100 family in malignant human breast-cancer cell lines and biopsy samples. Int J Cancer 1994; 57:684-690.

85. Kiyosawa K, Mokuno K, Murakami N et al. Cerebrospinal fluid 28-kDa calbindin as a possible marker for Purkinje cell damage. J Neurol Sci 1993; 118:29-33.

86. Usui A, Kato K, Murase M et al. Neural tissue-related proteins (NSE, G$^0\alpha$, 28-kDa calbindin, S100β and CK-BB) in serum and cerebrospinal fluid after cardiac arrest. J Neurol Sci 1994; 123:134-139.

87. Marshak DR, Pesce SA, Stanley LC et al. Increased S100β neurotrophic activity in Alzheimer's disease temporal lobe. Neurobiol of Aging 1991; 13:1-7.

88. van Eldik LJ, Griffin WST. S100β expression in Alzheimer's disease: relation to neuropathology in brain regions. Biochim Biophys Acta 1994; 1223:398-403.

89. Mattson MP, Barger SW, Cheng B et al. β-Amyloid precursor protein metabolites and loss of neuronal Ca^{2+} homeostasis in Alzheimer's disease. Trends Neurosci 1993; 16:409-414.

90. Gibson GE, Nielsen P, Sherman KA et al. Diminished mitogen-induced calcium uptake by lymphocytes from Alzheimer patients. Biol Psychiatry 1987; 22:1079-1086.

91. Petersen C, Goldman JE. Alterations in calcium content and biochemical processes in cultured skin fibroblasts from aged and Alzheimer donors. Proc Natl Acad Sci USA 1986; 83:2758-2762.

92. Grossmann A, Kukull WA, Jinneman JC et al. Intracellular calcium response is reduced in CD⁴⁺ lymphocytes in Alzheimer's disease and in older persons with Down's syndrome. Neurobiol Aging 1993; 14:177-185.

93. Föhr UG, Gibson GE, Tofel-Grehl B et al. Biochim Biophys Acta 1994; 1223:391-397.

94. Wu AHB, Valdes R Jr, Apple FS et al. Cardiac troponin-T immunoassay for diagnosis of acute myocardial infarction. Clin Chem 1994; 40:900-907.

95. Katus HA, Remppis A, Looser S et al. Enzyme-linked immunoassay of cardiac troponin T for the detection of acute myocardial infarction in patients. J Mol Cell Cardiol 1989; 21:1349-1353.

96. Katus HA, Looser S, Hallermeir K et al. Development and in vitro characterization of a new immunoassay for cardiac troponin T. Clin Chem 1992; 38:386-393.

97. Bodor GS, Porter S, Landt Y et al. Development of monoclonal antibodies for an assay of cardiac troponin-I and preliminary results in suspected cases of myocardial infarction. Clin Chem 1992; 38:2203-2214.

98. Thierfelder L, Watkins H, MacRae C et al. α-Tropomyosin and cardiac troponin T mutations cause familial hypertrophic cardiomyopathy: a disease of the sarcomere. Cell 1994; 77:701-712.

99. Hunt DF, Henderson RA, Shabanowitz J et al. Characterization of peptides bound to the class I MHC molecule HLA-A2.1 by mass spectrometry. Science 1992; 255:1261-1263.

100. Hauer CR, Staudenmann W, Kuster T et al. Protein sequence determination by ESI-MS and LSI-MS tandem mass spectrometry: parvalbumin primary structures from cat, gerbil and monkey skeletal muscle. Biochim Biophys Acta 1992; 1160:1-7.

101. Kuster T, Staudenmann W, Hughes GJ et al. Parvalbumin isoforms in chicken muscle and thymus. Amino acid sequence analysis of muscle parvalbumin by tandem mass spectrometry. Biochemistry 1991; 30:8812-8816.

102. Föhr UG, Weber BR, Müntener M et al. Human α and β parvalbumins. Eur J Biochem 1993; 215:719-727.

103. Hunt DF, Yates JR, Shabanowitz J et al. Amino acid sequence analysis of two mouse calbindin-D9k isoforms by tandem mass spectrometry. J Biol Chem 1989; 264:6580-6586.

104. Tsarbopoulos A, Gross M, Kumar R et al. Rapid identification of calbindin-D28k cyanogen bromide peptide fragments by plasma

desorption mass spectrometry. Biomed Environm Mass Spect 1989; 18:387-393.

105. Hall SC, Smith DM, Masiarz FR et al. Mass spectrometric and Edman sequencing of lipocortin I isolated by two-dimensional SDS/PAGE of human melanoma lysates. Proc Natl Acad Sci USA 1993; 90:1927-1931.

106. Sacks DB, Davis HW, Crimmins DL et al. Insulin-stimulated phosphorylation of calmodulin. Biochem J 1992; 286:211-216.

107. Zozulya S, Stryer L. Calcium-myristoyl protein switch. Proc Natl Acad Sci USA 1992; 89:11569-11573.

108. Dizhoor AM, Chen C-K, Olshevskaya E et al. Role of the acylated amino terminus of recoverin in Ca^{2+}-dependent membrane interaction. Science 1993; 259:829-831.

109. Pedrocchi M, Hauer CR, Schäfer BW et al. Analysis of Ca^{2+}-binding proteins in human heart by HPLC-electrospray mass spectrometry. Biochem Biophys Res Commun 1993; 197:529-535.

110. Cochran AJ, Lu H-F, Li P-X et al. S-100 protein remains a practical marker for melanocytic and other tumours. Melanoma Res 1993; 3:325-330.

111. Hanson CA, Bockenstedt PL, Schnitzer B et al. S100-positive, T-cell chronic lymphoproliferative disease: An aggressive disorder of an uncommon T-cell subset. Blood 1991; 78:1803-1813.

112. Oka K, Nakano T, Tatsuo A. Adenocarcinoma of the cervix treated with radiation alone: Prognostic significance of S-100 protein and vimentin immunostaining. Obstet Gynecol 1992; 79:347-350.

113. Marin F, Kovacs K, Stefaneanu L et al. S-100 protein immunopositivity in human nontumorous hypophyses and pituitary adenomas. Endocr Pathol 1992; 3:28-38.

114. Hachitanda Y, Nakagawara A, Nagoshi M et al. Prognostic value of N-myc oncogene amplification and S-100 protein positivity in children with neuroblastic tumors. Acta Pathol Jpn 1992; 42:639-644.

115. Gupta RK, Naran S, Dowle C et al. Coexpression of vimentin, cytokeratin and S-100 in monomorphic adenoma of salivary gland; value of marker studies in the differential diagnosis of salivary gland tumours. Cytopathology 1992; 3:303-309.

116. Engelkamp D, Schäfer BW, Erne P et al. S100α, CAPL, and CACY: Molecular cloning and expression analysis of three calcium-binding proteins from human heart. Biochemistry 1992; 31: 10258-10264.

117. Pedrocchi M, Schäfer BW, Durussel I et al. Purification and characterization of the recombinant human calcium-binding S100 proteins CAPL and CACY. Biochemistry 1994; 33:6732-6738.

INDEX

Page numbers in italics denote figures (f) or tables (t).

MOLECULAR BIOLOGY
INTELLIGENCE UNIT
AVAILABLE AND UPCOMING TITLES

NEUROSCIENCE INTELLIGENCE UNIT

AVAILABLE AND UPCOMING TITLES

☐ Neurodegenerative Diseases and Mitochondrial
Metabolism
M. Flint Beal, Harvard University

☐ Molecular and Cellular Mechanisms of Neostriatum
*Marjorie A. Ariano and D. James Surmeier,
Chicago Medical School*

☐ Ca^{2+} Regulation By Ca^{2+}-Binding Proteins
in Neurodegenerative Disorders
*Claus W. Heizmann and Katharina Braun,
University of Zurich, Federal Institute for Neurobiology,
Magdeburg*

☐ Measuring Movement and Locomotion: From
Invertebrates to Humans
*Klaus-Peter Ossenkopp, Martin Kavaliers and Paul
Sanberg, University of Western Ontario and
University of South Florida*

☐ Triple Repeats in Inherited Neurologic Disease
Henry Epstein, University of Texas-Houston

☐ Cholecystokinin and Anxiety
Jacques Bradwejn, McGill University

☐ Neurofilament Structure and Function
Gerry Shaw, University of Florida

☐ Molecular and Functional Biology
of Neurotropic Factors
Karoly Nikolics, Genentech

☐ Prion-related Encephalopathies: Molecular
Mechanisms
*Gianluigi Forloni, Istituto di Ricerche Farmacologiche
"Mario Negri"-Milan*

☐ Neurotoxins and Ion Channels
*Alan Harvey, A.J. Anderson and E.G. Rowan,
University of Strathclyde*

☐ Analysis and Modeling of the Mammalian Cortex
Malcolm P. Young, University of Oxford

☐ Free Radical Metabolism and Brain Dysfunction
Irène Ceballos-Picot, Hôpital Necker-Paris

☐ Molecular Mechanisms of the Action of Benzodiazepines
*Adam Doble and Ian L. Martin,
Rhône-Poulenc Rorer and University of Alberta*

☐ Neurodevelopmental Hypothesis of Schizophrenia
*John L. Waddington and Peter Buckley,
Royal College of Surgeons-Ireland*

☐ Synaptic Plasticity in the Retina
*H.J. Wagner, Mustafa Djamgoz and Reto Weiler,
University of Tübingen*

☐ Non-classical Properties of Acetylcholine
Margaret Appleyard, Royal Free Hospital-London

☐ Molecular Mechanisms of Segmental Patterning
in the Vertebrate Nervous System
*David G. Wilkinson,
National Institute of Medical Research-UK*

☐ Molecular Character of Memory in the Prefrontal Cortex
Fraser Wilson, Yale University

MEDICAL INTELLIGENCE UNIT

AVAILABLE AND UPCOMING TITLES

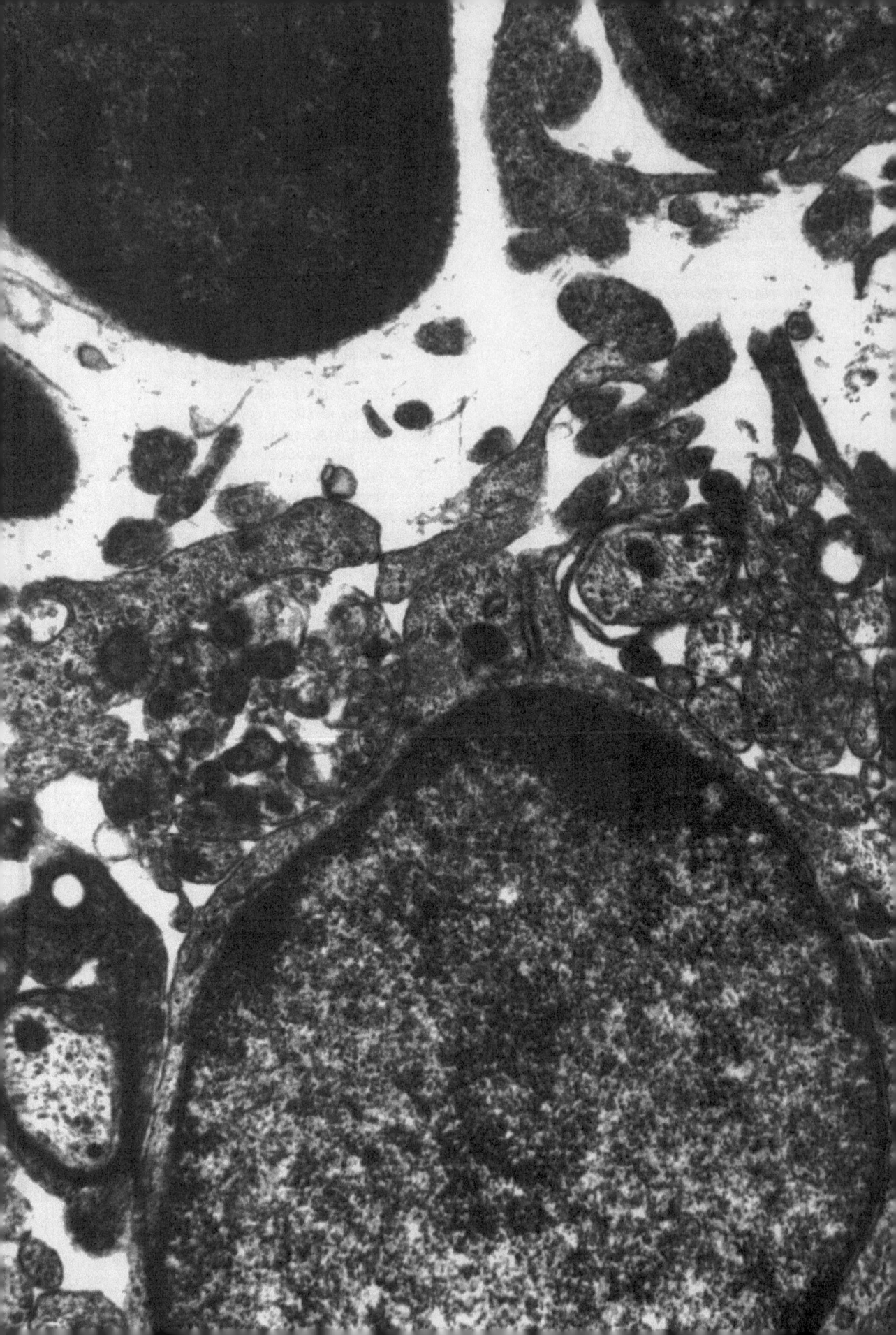

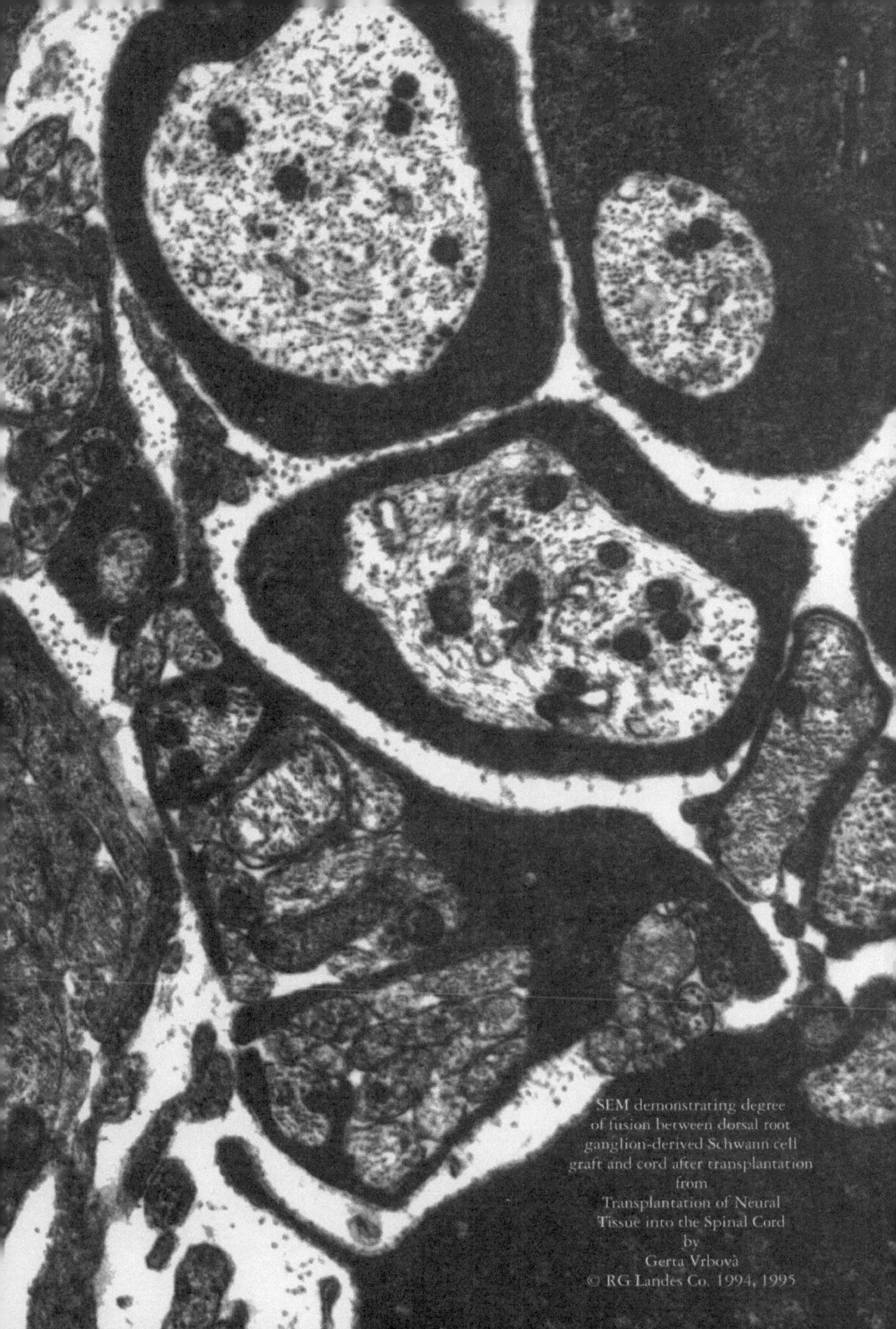

SEM demonstrating degree
of fusion between dorsal root
ganglion-derived Schwann cell
graft and cord after transplantation
from
Transplantation of Neural
Tissue into the Spinal Cord
by
Gerta Vrbová
© RG Landes Co. 1994, 1995